Experiments in Engineering Physics

Dr. J. Anjaiah

Experiments in Engineering Physics

J. Anjaiah, Ph.D.
The University of Dodoma, Tanzania
&
Professor of Physics
Geethanjali College of Engineering & Technology
Telangana, India

Lulu Publications, USA

Copyright © 2015 by J. Anjaiah

All rights reserved. This book or any portion thereof may not be reproduced or used in any manner whatsoever without the express written permission of the publisher except for the use of brief quotations in a book review or scholarly journal.

First Printing: 2015

ISBN: 978-1-312-88896-8

Lulu Publications, USA.

Preface

This book is intended for the first year engineering graduate students of Jawaharlal Nehru Technological University-Hyderabad, India. The material in this book could be covered the laboratory experiments in Engineering Physics course of Bachelor of Technology (B.Tech.) programme. There are many books as lab manuals for the first year students are available today each with its own particular approach. The approach taken in this book is to present to the beginning student an extensive easy approach of introductory levels, those who are coming from their intermediate studies. Special emphasis was placed on presenting the basic concepts and results. Two additional experiments are also included apart from the 13 experiments from their regular syllabus. At the end of each experiment viva questions are included.

I express my heartfelt thanks to our secretary, Mr. G.R. Ravinder Reddy, Geethanjali group of institutions, India for his encouragement. I also express my sincere thanks to authorities of University of Dodoma, Tanzania for their support.

Finally I thank the Lulu publishers for publishing this book.

Instructions for the laboratory

- *The experiments are designed to illustrate phenomena in different areas of Physics and to expose you to measuring instruments.*
- *Conduct the experiments with interest and an attitude of learning.*
- *Be honest in recording and representing your data. Never make up readings to get a better fit for a graph. If a particular reading appears wrong repeat the measurement carefully.*
- *All presentations of data, tables and graphs calculations should be neatly and carefully done.*
- *Learn to optimize on usage of graph papers. Graphs should be neatly drawn with pencil. Always label graphs and the axes and display units.*
- *Handle instruments with care.*

Contents

Preface i

Instructions for the laboratory ii

1. Dispersive power of the material of a prism-spectrometer 1
2. Determination of wavelength of a source-diffraction grating 9
3. Newton's rings 15
4. Melde's experiment 21
5. Time constant of an R-C circuit 27
6. L-C-R circuit 31
7. Magnetic field along the axis of current carrying coil 39
8. Study the characteristics of LED and LASER sources 45
9. Numerical aperture and bending losses in a given optical fibre 53
10. Energy gap of a material of P-N junction 59
11. Torsional pendulum 65
12. Wavelength of a LASER light using diffraction grating 71
13. Characteristics of a solar cell. 75
14. Hysteresis loop of a ferromagnetic material 81
15. Volume resonator 89

References 93

Graphs 94

1. Dispersive power of the material of a prism

Aim: To determine the dispersive power of the material of a prism by using spectrometer.

Apparatus:
1) Spectrometer
2) Mercury vapour lamp
3) Flint glass prism
4) Reading lens

Principle: The dispersive power of the material of the given prism is expressed by the relation,

$$\omega = \frac{\mu_B - \mu_R}{\mu - 1}$$

where 'μ_R' and 'μ_B' are the refractive indices of any two colours, but usually the red and blue colours are chosen to identify easily; and $\mu = \frac{\mu_B + \mu_R}{2}$.

Description:

The dispersive power of the material of a prism is determined by using spectrometer, it mainly consists of
(1) Collimator
(2) Telescope
(3) Prism table
(4) Circular scale and vernier.

i) The collimator: The collimator consists of a convex lens fitted to the inner end of a hollow tube, fixed to the instrument. Another hollow tube which exactly fits into the fixed tube and can be moved in or out by means of a rack and pinion screw. The collimator is fixed to the instrument and cannot be rotated. The collimator is used to obtain a parallel beam of light from a given source.

ii) The Telescope: This is an astronomical telescope whose objective is fitted to the inner end of a hollow tube. In this tube there is another hollow tube which can be moved. At the outer end, the tube carries the Ramsden's eye piece with cross-wires. The distance of the cross-wires from the eye-piece can be altered by pushing in or drawing out the eye-piece.

The telescope can be turned about an axis coinciding with the axis of rotation of the prism table and can be clamped on any position with the given screw. The angle of rotation can be measured, on a circular scale which is fixed to the telescope and moves along with it. The telescope is used to receive the parallel beam of light from the collimator.

iii) Prism table: It is a small circular table provided with three leveling screws and is used for keeping the prism on it. The prism table can be raised or lowered and clamped in any position by a screw. By means of another screw it can be fixed to the vernier table and the two will then turn together.

iv) The circular scale: This is a circular metal plate attached to the telescope and rotates with it.

Determination of angle of the prism (A):
Procedure:

The primary adjustments of the spectrometer are to be done. The prism is placed at the centre of the prism table such that both the refracting edges of the prism are facing the collimator symmetrically. Then the prism is fixed. The telescope is released and rotated to observe reflected image of the prism from one face. Adjust upto the reflected image coincides with the cross wires. The readings of the two verniers are to be noted. The telescope is rotated such that the reflected image of the second face is focused. Then readings of the both the verniers are to be noted. Then the difference of the both reading gives the value of $2A$ from with the refracting angle can be calculated.

Determination of angle of minimum deviation (D_m):

The vernier table is clamped and the prism table is released. The prism is clamped centrally on the prism table such that the surface of the ground glass is almost parallel to the axis of the collimator and the light from collimator incident or the polished surface of the prism emerges out from the other polished surface.

Looking at the image, the prism table is slowly turned such that the image moves towards the direct position. The telescope is also moved so as to keep the image of the slit in the field of view. At certain stage it will be found that the image changes its direction of motion even though the prism is continued to move in the same direction.

The position of the prism is fixed when refracted image of the slit just retraces its path, which is the position of minimum deviation. The telescope is focused such that the image coincides with the vertical cross wires. The readings of two verniers are noted. The prism is removed and the telescope is rotated such that the direct image of the slit coincides with cross wires. Readings of two verniers are noted. The difference between the respective readings of the verniers gives the angle of minimum deviation of the prism (D_m), and then the readings are tabulated.

Direct readings: V_1=_____

V_2=_____

Colour	Reading of spectrometer						Difference in (2A) Vernier readings			Angle of minimum deviation (D_m)
	V'_1			V'_2					Mean (θ)	
	MSR	VC	Total	MSR	VC	Total	$V'_1 \sim V_1$	$V'_2 \sim V_2$		
Blue										
Red										

Refractive index of the given prism can be calculated by using the formula,

$$\mu = \frac{\sin\frac{A+D_m}{2}}{\sin\frac{A}{2}}$$

Dispersive power of the prism:

The usual adjustments of the spectrometer are made. The refractive angle of the prism is found, and then the prism is mounted on the prism table such that the light from the mercury vapour lamp is deserved through the spectrometer the spectrometer is observed through the telescope. The prism is deserved and set in minimum deviation position for the violet line and the vertical cross wires are made to coincide with green line. The readings on the vernier scales are to be noted. The direct reading when the vertical cross wire coincides with the slit is also be noted. The difference between these two readings gives the angle of minimum deviation of violet.

The experiment is repeated for different lines in the spectrum and the readings are to be tabulated. By knowing the values for any two colours, the dispersive power of the material of the prism is determined from this formula,

$$\omega = \frac{\mu_B - \mu_R}{\mu - 1}$$

Precautions:
1. The prism should be adjusted for each colour separately.
2. The optical adjustments must be done carefully before starting the experiment.
3. The prism must be set symmetrically on the prism table.
4. Reading on both verniers are to be taken.
5. The polished surfaces of the prism should not be touched. It should be handled by its edges.

Result:
Dispersive power of the material of the prism is _____

Viva questions

1. *Explain dispersive power of a prism.*
A. The separation of white light into its constituent colours is called dispersion of light. When the light under goes dispersion, the band of colours obtained is known as spectrum.

   Angular dispersion of two colours is the difference between the angles of deviation of two colors. The angular dispersion of red and violet is equal to $\delta_V - \delta_R$.

   The ratio of angular dispersion of two extreme colours to their mean deviation is known as the dispersive power of the material of a prism and it is denoted by ω.

   $$\omega = \frac{Angular\ dispersion\ between\ red\ and\ violet}{Their\ mean\ deviation}$$

   $$\omega = \frac{\delta_V - \delta_R}{\frac{\delta_V + \delta_R}{2}} \quad let \quad \frac{\delta_V + \delta_R}{2} = \delta$$

   $$\omega = \frac{\delta_V - \delta_R}{\delta - 1}$$

2. *What is the Ramsden's eye piece?*
A.
   a) Its field lens and eye lens are of same focal length. They have separated by a distance of 2/3f.
   b) Curved surfaces face each other.
   c) The image due to objective is at a distance of f/4 from the eye lens in front of it. So, it is called positive eye piece.
   d) It reduces spherical aberration. By making each lens an achromatic doublet, chromatic aberration can be reduced.
   e) It can be used for measurements.

3. *Explain the working of Huygens's eye piece.*
A.
   a) Its field lens is of focal length 3f and eye lens is of focal length 'f'. They are separated by a distance 2f.
   b) Curved surfaces of field lens face the objective and the curved surface of eye lens faces the field lens.
   c) The image due to objective would have formed at a distance of 3f/2 in the absence of field lens. The image is behind the field lens. So it is called negative eye piece.
   d) It minimizes spherical and chromatic aberrations.

e) It can't be used for measurements.
4. *Explain about different types of aberrations.*
A. The defects and imperfections developed in images are known as abberations.
   i) *Chromatic aberration*
      The aberration formed a lens with white light or composite light producing colored images are called chromatic aberrations.
   ii) *Spherical aberration*
      The formation of line image along the principle axis when a point object is placed on the principle axis is called spherical aberration. It is due to large aperture of lens.

   Monochromatic aberrations are of five types
   a) Spherical aberration
   b) Coma
   c) Astigmatism
   d) Curvature
   e) Distortion

5. *Define dispersive power of a prism*
A. Dispersive power of a prism indicates the ability of the material of the prism to disperse the light rays. It is defined as the angular dispersion to the deviation of the mean ray.

6. *How does refractive index change with wave length?*
A. Higher the wave length, smaller is the refractive index.

7. *Does the deviation depend on the angle of prism?*
A. Yes, greater the angle of the prism, more is the deviation

8. *What is a prism?*
A. A transparent medium like glass bounded by two smooth surfaces which are transparent and one rough surface which is not transparent.

9. *What is meant by angular dispersion?*
A. It is the difference in angular deviation between any two colours.

10. *What is refractive index?*

A. The ratio of sine of angle of incidence in the first medium to the sine of angle of refraction in the second medium.

11. What is Spectrometer?
A. It is an optical instrument which is used to study the nature of light. It consists of collimator, prism table and telescope.

12. What is the function of a collimator?
A. It will produce parallel beam of light.

13. What do you mean by angle of prism?
A. Angle between two refracting surfaces of the prism.

14. What is dispersion of light?
A. When the light is allowed to fall on one of the refracting surfaces of a prism, it is split into its constituent colours. This splitting of light into its constituent colours by refraction through prism is called dispersion of light.

15. What is the main optical action of the prism?
A. The main optical action of a prism is to disperse white light into its component parts. Dispersion of light is minor optical action of prism, but main effect of a prism is to deviate a beam of light.

16. What type of prism do you use in this experiment?
A. Crown prism.

17. What are the units of dispersive power?
A. No units.

18. What type of light do you use in this experiment?
A. White light.

19. Which colour in the spectrum is having more refractive index?
A. Violet colour. The refractive index for violet is maximum when compared to other colours. The refractive index for red is minimum when compared to other colours.

2. Determination of wavelength of a source - diffraction grating

Aim: To determine the wavelength of a given source of light by using the diffraction grating with (a) normal incidence and (b) minimum deviation method.

Apparatus: Plane diffraction grating, spectrometer, spirit level, reading lens, sodium vapour lamp.

Formula:

$\lambda = \dfrac{\sin\theta}{Nn}$ Å .... (Normal incidence method)

$\lambda = \dfrac{2\sin(D_m/2)}{Nn}$ Å .... (Minimum deviation method)

Description:

A plane diffraction grating consists of a parallel-sided glass plate with equidistant and fine parallel lines drawn very closely upon by means of a diamond point. The number of lines drawn is about 15,000 per inch (the gratings used in the laboratory are exact replicas of the original gratings on cellulose film).

Theory:

When light of wavelength 'λ' is incident normally on a diffraction grating having 'N' lines per inch, and if 'θ' is the angle of diffraction in the nth order spectrum, then

$$nN\lambda = \sin\theta$$

$$\text{(or) } \lambda = \dfrac{\sin\theta}{Nn}$$

$$\lambda = \left(\dfrac{\sin\theta \times 2.54}{n \times 15{,}000}\right) \text{ cm}$$

$N = 15{,}000$ lines/inch (or)

$N = \dfrac{15{,}000}{2.54}$ lines/ cm

Again, when a parallel beam of monochromatic light incident upon a grating is diffracted in such a way that the angle of deviation is minimum then the wavelength of radiation is given by,

$$\lambda = \dfrac{2\sin(D_m/2)}{Nn}$$

$$\text{(or) } \lambda = \left(\dfrac{2 \times 2.54 \times \sin(D_m/2)}{15{,}000}\right)$$

where D_m = Angle of minimum deviation

Procedure:

The usual initial adjustments of the spectrometer are done. The least count of the vernier of the spectrometer is found.

a) Normal incidence:

The slit of the spectrometer is illuminated with sodium vapour lamp. The telescope is placed in line with the axis of the collimator and the direct image of the slit is observed. The slit is narrowed and the vertical cross-wire is made to coincide with the center of the image of the slit. The reading of one of the verniers is noted. The prism table is clamped firmly and the telescope is turned through exactly 90^0 and fixed in position.

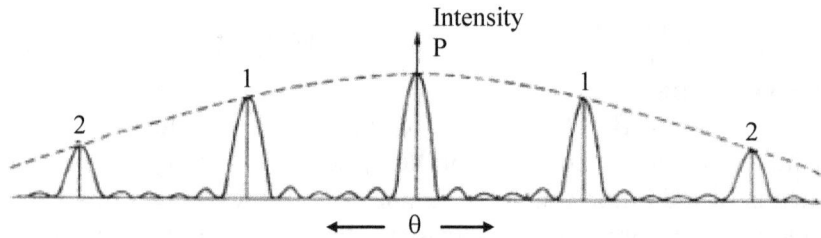

Fig. 5.1 The variation of resultant intensity

The grating is held with the rulings vertical and mounted in its holder on the prism table such that the plane of the grating passes through the center of the table and the ruled surface towards the collimator. The prism table is released and rotated until the image of the slit is seen in the telescope by reflection on the ruled side of the grating. The prism table is fixed after adjusting the point of intersection of the cross wires is on the image of the slit. Then the vernier table is released and rotated through exactly 45^0 from this position so that the ruled side of the grating faces the collimator. The vernier table is fixed in this position and the telescope is brought back to the direct reading position. Now the light from the collimator strikes the grating normally.

Measurement of λ:

The telescope is rotated so as to catch the first order diffracted image on one side, say on left (see fig). With sodium light two images of slit, very close to each other, can be seen. They are called D_1 and D_2 lines. The point of intersection of cross wires is set on the D_1 line and its reading is noted on both the verniers. Similarly the reading corresponding to the D_2 line is noted. Then telescope is turned to the other side i.e., right side and similarly the readings corresponding to D_1 and D_2 lines of the first order spectrum are noted. Half the difference in the readings corresponding to any one line gives the angle of diffraction (θ) for that line in the first order spectrum.

The experiment is repeated for the second order spectrum.

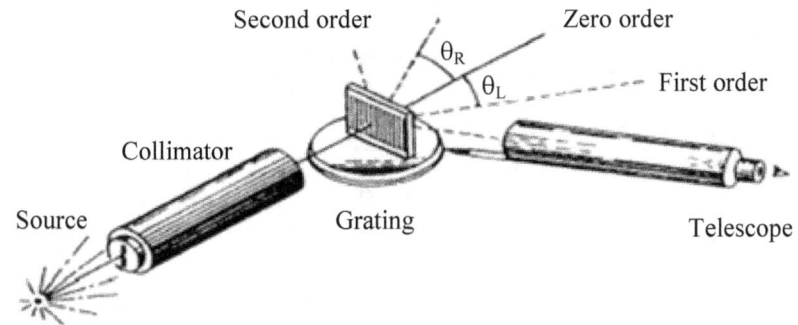

Fig. 5.2 Diffraction through a grating

Observations:
Direct ray readings: $V_1 =$ _____ $V_2 =$ _____
Direct ray readings: $V_1 =$ _____ $V_2 =$ _____

b) Minimum deviation method:
The direct image of the slit is observed through the telescope. The point of intersection of the cross wires is set on the sharp image of the slit. The vernier table is fixed and the reading on the circular scale is noted.

The prism table is released from the vernier table. The telescope is turned to one side, (say right) and the first order diffracted image is observed. The prism table is slowly rotated to the right. As it is slowly rotated towards right side, the image first moves towards left, reaches a limiting position and then retraces its path. In this limiting position, the telescope is fixed such that the point of intersection of the cross wires is on the D_1 line and the reading on the circular scale is taken. The difference between the direct reading and this reading gives the angle of minimum deviation for the line D_1 in the first order spectrum. Similarly the angle of minimum deviation for the D_2 line of the first order found.

Normal incidence method

Order of the spectrum	Line	Reading of the spectrometer								Angle of diffraction = Direct ray reading-Reading of the spectrometer			$\lambda = \dfrac{\sin\theta}{Nn}$
		V'_1			V'_2					$V'_1 \sim V_1$	$V'_2 \sim V_2$	Mean (θ)	
		MSR	VC	Total	MSR	VC	Total						
First order	D_1												
	D_2												
Second order	D_1												
	D_2												

Minimum deviation method

Order of the spectrum	Line	Reading of the spectrometer								Angle of diffraction = Direct ray reading-Reading of the spectrometer			$\lambda = \dfrac{2\sin\frac{D_m}{2}}{Nn}$
		V'_1			V'_2					$V'_1 \sim V_1$	$V'_2 \sim V_2$	Mean (θ)	
		MSR	VC	Total	MSR	VC	Total						
First order	D_1												
	D_2												
Second order	D_1												
	D_2												

Next, the angle of minimum deviation for D_1 and D_2 lines in the second order spectrum is found similarly. The results are tabulated in the table above.

Precautions:
1. Always the grating should be held by the edges. The ruled surface should not be touched.
2. Light from the collimator should be uniformly incident on the entire surface of the grating.

Result :
   The wave length of a given source of light is _____ A^0

Viva questions

1. What is plane transmission diffraction grating?
A. Plane transmission diffraction grating is an optically plane parallel glass plate on which equidistant, extremely close grooves are made by ruling with a diamond point.

2. In our experiment, what class of diffraction does occur and how?
A. Fraunhofer class of diffraction occurs. Since the spectrometer is focused for parallel rays, the source and the image are effectively at infinite distances from the grating.

3. How are the commercial gratings are made?
A. Commercial gratings are made by pouring properly diluted cellulose acetate on the actual grating and drying it to a thin strong film. The film is detached from the original grating and is mounted between two glass plates. A commercial grating is called replica grating. In our experiment we use plane type replica grating.

3. Newton's rings

Aim: To determine the radius of curvature of given plano convex lens by using Newton's rings method.

Apparatus: Plano convex lens of large radius of curvature, a perfectly plane glass plate, black cloth or paper, travelling microscope, reading less, sodium vapour lamp, reflecting glass plate fixed to a small stand.

Formula:

$$R = \frac{D_n^2 - D_m^2}{4\lambda(n-m)} \text{ cm} \quad \ldots\ldots\ldots\ldots\ldots \text{ (experimental)}$$

$$R = \frac{l^2}{6h} + \frac{h}{2} \text{ cm} \quad \ldots\ldots\ldots\ldots\ldots \text{ (theoretical)}$$

Description:

Combination of a plano-convex lens with its curved surface on the plane glass plate is placed on the platform of the travelling microscope. A black paper is placed under the glass plate. The reflecting glass plate is arranged to reflect the light from the sodium vapour lamp onto the combination. The microscope is focused on the combination. Alternate dark and bright rings with a dark central spots are observed. These are Newton's rings.

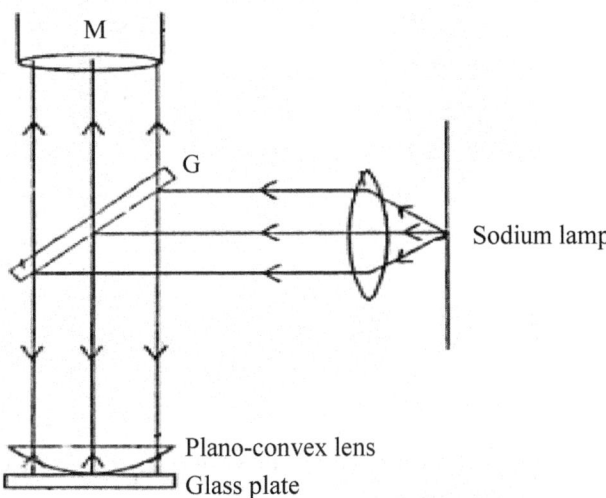

Fig. 6.1 Reflection and transmission of light in Newton's rings apparatus

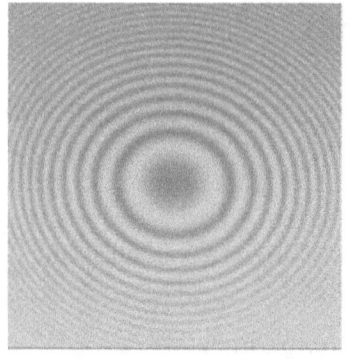

Fig. 6.2 Image of the rings   Fig. 6.3 Newton's rings apparatus

Procedure:

The microscope is moved to one side counting the number of rings. At the 12th ring, it is turned back. The vertical cross wire is made tangential to the 12th, 10th, 8th, 6th, 4th and 2nd rings and the readings are noted.

Similarly, the readings are noted when the cross wire is tangential to 2nd, 4th, 6th, 8th, 10th and 12th rings on the other side. The readings are tabulated in the table.

SNo.	Number of the ring	Microscope reading		Diameter of Ring $D=(a\sim b)$	D^2
		Left (a)	Right (b)		
1	12th				
2	10th				
3	8th				
4	6th				
5	4th				
6	2nd				

The radius of curvature of plano-convex is found by using spherometer from the formula.

$$R = \frac{l^2}{6h} + \frac{h}{2}$$

where, l = Mean distance between the two legs of the spherometer

h = Difference of readings of spherometer when the tip of the screw touches the curved surface of the lens and plane glass plate.

A graph is drawn between number of ring and D^2. It will be a straight line passing through the origin. From the graph, the values of D_n^2 and D_m^2 corresponding to n^{th} and m^{th} rings are found.

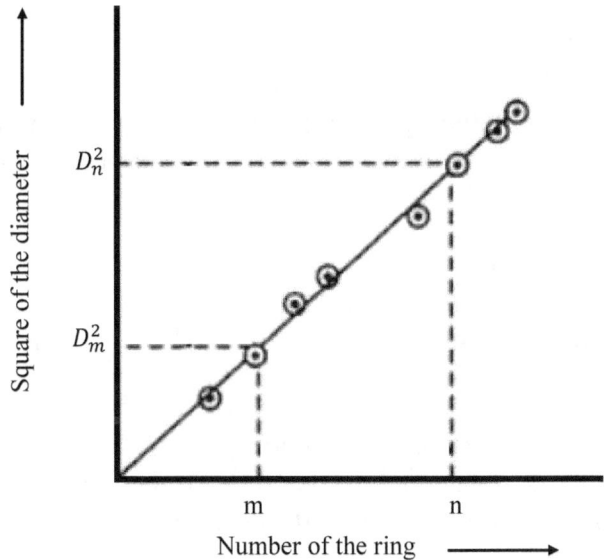

Fig. 6.4 Graph between the number of ring and D^2

The wavelength 'λ' of sodium light is found by formula,
$$R = \frac{D_n^2 - D_m^2}{4\lambda(n-m)} \text{ cm}$$
'λ' is wavelength of Sodium vapour lamp = 5893 Å

Precautions:
1. While taking observations, the microscope should be moved only in one direction to avoid the error due to back-lash.
2. The lens and glass plate must be perfectly clean.
3. The slow motion screw of the microscope must be used while taking readings.
4. The central spot must be dark.

Result:
The radius of curvature of plano-convex lens is ------------cm

Viva questions

1. What are Newton's rings?

A: When a plano-convex lens is placed on a glass plate an air film of gradually increasing thickness is formed between glass plate and plano-convex lens. When monochromatic light is allowed to fall normally on air film and viewed in reflected light, alternate dark and bright rings formed are known as Newton's rings.

2. How is Newton's rings formed.

A: These rings are formed as a result of interference between the light waves reflected from the upper and lower surface of the air film developed between the convex surface of the plano-convex lens and plane glass plate.

3. Why the Newton's rings are circular?

A: They are circular because the air film formed is wedge shaped and locus of the points of equal thickness is circles concentric with the point of contact.

4. What is the function of 45^0 inclined plane glass plate?

A: The transparent glass plate inclined at 45^0 turns the light rays coming from an extended source to 90^0 and, so the rays fall normally on plano-convex lens.

Define the following terms.

i). *Diffraction:* The bending of the wave front or its deviation from its original direction when it meets a small obstacle or at its edge known as diffraction.

ii). *Interference:* The variation in intensity in the region of superposition of two or more waves of same amplitude and frequency whose phase relationship does not change with time is known as interference.

iii). *Polarization:* It is the property of getting one sidedness in the direction of vibrations of electric vector of light wave.

iv). *Principle of superposition:* The principle of superposition of waves states that when two or more waves travel through the same position of a medium simultaneously, the resultant displacement at any point is the vector sum of the displacement due to individual waves.

v). *Constructive super position of waves:* When two sources vibrate with same amplitude such that crest falls on crest or trough falls on the trough, a maximum displacement is produced. When such maximum displacement occurs the waves are said to superimpose constructively.

vi). *Destructive superposition:* When two sources vibrate with the same amplitude such that crest falls on trough or vice versa, a minimum displacement is produced. When such minimum displacement occurs the waves are said to superimpose destructively.

4. Melde's experiment

Aim: To determine the frequency of a tuning fork or vibrator, using Melde's arrangement.

Apparatus: Electrically driven tuning fork or vibrator, thread, a small light pan, weight box, smooth frictionless pulley fixed to stand, battery and connecting wires.

Formula:

$$n = \frac{1}{2l}\sqrt{T/\mu} \quad \ldots\ldots\ldots\ldots \text{(Transverse mode)}$$

$$n = \frac{1}{l}\sqrt{T/\mu} \quad \ldots\ldots\ldots\ldots \text{(Longitudinal mode)}$$

Description:

This consists of a tuning fork of low frequency mounted on a heavy base. An electromagnet is arranged between the prongs of fork without touching it. A small light spring is attached to one of the prongs. Screw just makes contact with the spring, through a rheostat and plug key. When the circuit is closed, the electromagnet is energized and the prongs are pulled together. The spring loses contact with the circuit, the current momentarily stops, the electromagnet loses its magnetism and the prongs go back to their original position restoring contact of the spring with the screw and the circuit is closed and so on. This fork or vibrator is maintained in oscillation continuously. One end of the thread is attached to the prong and the other end passes over a smooth pulley. The other end of the thread carries the small pan. The thread must be horizontal.

Theory:

When the fork vibrates, the disturbance travels along the thread. It undergoes reflection at the pulley. The reflected wave is superimposed on the direct wave and standing waves are formed. The thread is divided into a number of loops. By adjusting the length of the thread and weight in the scale pan the loops can be well defined.

Suppose the thread is arranged so that the prongs vibrate parallel to the length of the thread. Then the arrangement is called longitudinal position. In this method the frequency of the thread is half the frequency of the fork. If n is the frequency of thread, the frequency of fork is 2n.

Suppose the thread is arranged so that the prongs vibrate perpendicular to the length of the thread. Then the arrangement is called transverse position.

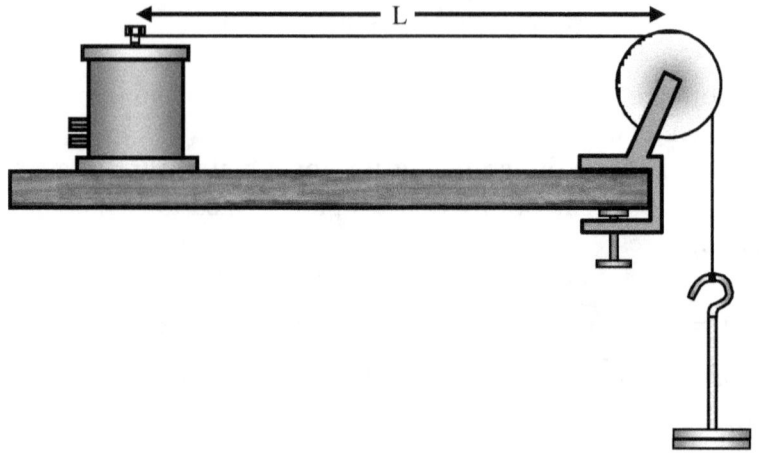

Fig. 3.1 Melde's experimental setup

In this method, the frequency of thread is equal to the frequency of the fork. If n is the frequency of thread, the frequency of fork is n.

If '*l*' is length of the thread between the prong and the pulley and if the thread breaks up into 'X' well defined loops, then the frequency of thread is given by

$$n = \frac{1}{2l}\sqrt{T/\mu}$$

where 'μ' is mass per unit lengh of the thread (linear density).
  '*l*' is length of a single loop.
  '*T*' is the tension applied = $(M+P)g$
  '*g*' is acceleration due to gravity at the place.(980 cm/s^2)
  '*P*' is mass of pan and
  '*M*' is load added in the pan.

Procedure:
Transverse position:

SNo.	Load in the pan M	Tension T=(M+P)g dynes	No. of loops (x)	Length of x Loops (L)	Length of each Loop l=L/x	$\sqrt{T/l}$
1	20 gm					
2	40 gm					
3	60 gm					
4	80 gm					

The apparatus is arranged in the transverse position and the experiment is reaped in the same manner as above. The readings are tabulated in the above table.

The frequency of tuning fork $(n) = \dfrac{1}{2\sqrt{\mu}} \dfrac{\sqrt{T}}{l} = $ _____ Hz

Longitudinal position:

The apparatus is arranged in the longitudinal position so that the prongs vibrate parallel to the thread. The thread is made horizontal. A small load is placed in the pan. The circuit is closed. The length of the thread is adjusted by moving the pulley to and fro slightly. If necessary the load in the pan is slightly adjusted until the thread breaks into a number of well defined loops. The experiment is repeated for different tensions and results are tabulated in the table:

SNo.	Load in the pan M	Tension T=(M+P)g dynes	No. of loops (x)	Length of x Loops (L)	Length of each Loop l=L/x	$\sqrt{T/l}$
1	20 gm					
2	40 gm					
3	60 gm					
4	80 gm					

The frequency of tuning fork $(n) = \dfrac{1}{\sqrt{\mu}} \dfrac{\sqrt{T}}{l} = $ _____ Hz

Observations:
Linear density of the thread $(\mu) =$

$\dfrac{\text{Mass of the string (thread)}}{\text{Length of the string}} = $ _____ gm/cm.

Mass of the pan, 'P' = ____ gm.

Precautions:
1. The loops must be well-defined.
2. The plane of vibration of the thread must be vertical
3. In counting the loops, the loops at the two extreme ends must not be taken into account.

Result:
The frequency of tuning fork in transverse mode = _____ Hz
The frequency of tuning fork in longitudinal mode = _____ Hz

Viva questions

Define and explain the following terms.

1. *Audible frequency range:* The range of frequencies of sounds which are between the 20 *Hz* and 20,000 *Hz* that produce the sensation of hearing. Examples for these are bomb blasts, etc. they effect the eardrum causing acute pain.

2. *Infrasonic sounds:* Sounds which have the frequencies below 20*Hz*. Examples for these are waves of earth quakes, volcanoes etc they cause damage to the human body.

3. *Time period (T):* It is the time taken by the disturbance to advance through a distance of one wavelength in the direction of propagation. It is equal to the period of oscillation of any particle on the wave. Unit for this is *second*.

4. *Amplitude (A):* The maximum displacement of any particle on the curve from its mean position in either direction. Unit for this is *meter*

5. *Frequency (v):* Frequency is equal to number of vibrations per second completed by any particle on the curve. Or the number of wavelengths that take place in unit time along the direction of propagation of the wave. Its units are *Hertz*.

6. *Wavelength (λ):* The distance between two successive particles which are in the same phase of vibration on the wave is known as wavelength.
   S I unit: *meter*
   In transverse wave, the distance between two successive crests or trough is known as wavelength. In longitudinal wave, the distance between two successive compression and rarefactions is known as wavelength.

7. *Phase (Φ):* The phase of the wave at any instant at a point is the state of vibration of the particle at that point, with regards to its position and direction. The S.I. unit of phase is *radian*.

8. *Wave motion:* The process of transmitting energy through the vibrations of the particles of the medium is known as wave motion.
(or)
Wave motion can be defined as a form of disturbance which travels through the medium due to repeated periodic motion of the particles of medium about their mean positions.

9. *Progressive waves:* A wave which travels from a point into an infinite medium and never returns to the origin is called a progressive wave.

10. *Stationary wave:* When two progressive waves of same amplitude and wavelength travelling through a medium in opposite directions are super imposed, then the waves formed are known as stationary waves.

11. *Transverse wave:* If the direction of propagation of the waves is perpendicular to the direction of vibration of the particles, the wave are said to be transverse wave.

12. *Longitudinal waves:* If the particles of medium vibrate parallel to the direction of the propagation of waves, then the waves are said to be longitudinal waves.

13. *Resonance:* If a body vibrates under the influence of a periodic force impressed on it, and if the natural frequency of the body coincides with the frequency of the periodic force on it, the body vibrates with increasing amplitude. This phenomenon is known as "Resonance".

Ex: i) consider a child in the swing on giving a periodic push to the swing in the same direction every time it passes the extreme position, the amplitude of the swing goes on increasing and soon becomes very large. Since the natural frequency of the swing is equal to the frequency of the periodic force, the resonance occurs between the two and hence the amplitude increases.

ii) Resonance is always not desirable. When a band of soldiers are marching on a bridge, they are commanded to go out of step. If they were to go marching in step, and if the natural frequency of the bridge happens to be equal to the frequency of the step in marching, resonance occurs between the two. Due to the resonance, the amplitude of vibrations may increase enormously. The bridge may vibrate violently and collapse. Hence they are asked to go out of step.

14. *Doppler Effect:* The phenomenon of apparent change in the frequency of a sound wave due to the relative motion between the source and the listener.

5. Time constant of an R-C circuit

Aim: To study the growth and decay of charge in an R-C Circuit and determine the value of time constant.

Apparatus: Source of e.m.f (battery eliminator), resistors, capacitors, voltmeter, galvanometer, stop clock, tap key and connecting wires.

Formula:
Time constant = RC .... (Theoretical)

Theory:
When a condenser C is charged through a resistance R, the charge in condenser increases with time as an exponential function. If q is the instantaneous charge at time t, we have,

$$q = q_0(1 - e^{-t/RC})$$

Where, 'q' is the charge and 't' is the time at that instant, 'q_0' is the maximum charge

The product 'RC' is called time constant of the circuit. It is the time taken to establish

$(1 - e^{-\frac{t}{RC}})$ part of the maximum charge in the condenser. It is equal to the time taken to establish 0.632 part of the total charge. When a condenser is discharged through a resistance, the charge falls in accordance with the formula

$$q = q_0 e^{-t/RC}$$

The time constant in this case is equal to the time, taken to decrease the charge of 'e' part of the maximum charge. It is equal to the time taken to discharge to a value of 0.368 part of maximum charge.

Thus we can observe that: smaller is the time constant, more rapid is the discharge of the capacitor.

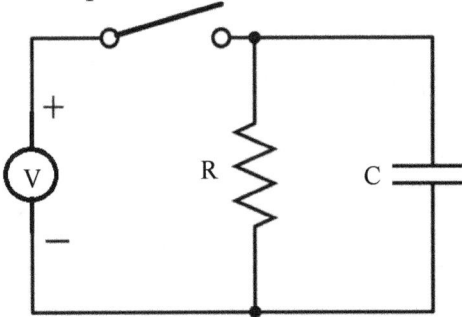

Fig.8.1. Circuit diagram of an R-C circuit

Procedure:

The circuit is connected as shown in figure, taking one set of R and C. The capacitor 'C' is charged for a short time till the deflection in the galvanometer is maximum, but within the scale. The deflections are noted for every five seconds. The tap key is then released. The capacitor now starts discharging through the resistor R. The deflection decreases steadily. The stop clock is started and readings are noted down for every five seconds.

The experiment is repeated for the other sets of R and C and observations are tabulated in the table below:

Observation Table:

Time (in seconds)	$R_1 C_1$ Combination $R_1 =$ ____ Ω $C_1 =$ ____ μF		$R_2 C_2$ Combination $R_2 =$ ____ Ω $C_2 =$ ____ μF	
	Charging	Discharging	Charging	Discharging
5				
10				
15				
20				
25				
30				
35				
40				
45				
50				
55				
60				
65				
70				
75				

Graph:
The graph is drawn between the time on x-axis and deflections on the y-axis.

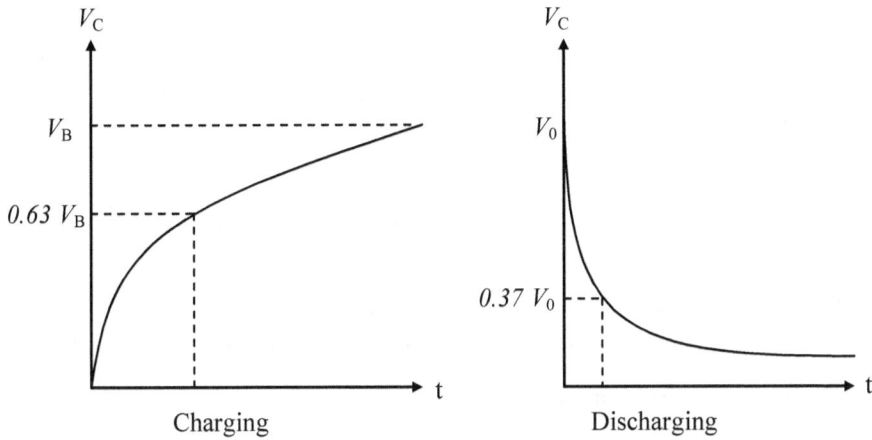

Fig. 8.2 Graph between the time and voltage

From this plot find the time for which the deflection falls to 0.368 of starting value in the case of discharging. It is equal to time constant.

Precautions:
1. There should not be any leakage of charge.
2. Connecting wires used to the condenser must be short.

Result:
Time Constant of R_1C_1 is _____Seconds. (Theoretical)

Time Constant of R_1C_1 is _____Seconds. (Experimental)

Time Constant of R_2C_2 is _____Seconds. (Theoretical)

Time Constant of R_2C_2 is _____Seconds. (Experimental)

Viva questions

1. *On what factors does the rate of charging and discharging of a capacitor depend?*
A. Capacity of capacitor and resistance of the resister used in the circuit.

2. *Why a high resistance is used in this experiment?*
A. If high resistance is used in the circuit, then the capacitive time constant becomes sufficiently high.

3. *Define the 'time constant' of the R-C circuit.*
A. The time constant of the R-C circuit during both charging and discharging processes of a capacitor is equal to the product of the magnitudes of both resistance R and the capacitance C of the capacitor.
i.e., $t_c = R \times C$

4. *How does the circuit vary with time during discharging in a R-C circuit?*
A. The current at any time 't' during discharging is given by $I = I_0 e^{\frac{-t}{RC}}$ where I_0 is the maximum value of the current and time constant.

5. *What is a blocking capacitor?*
A. The capacitor that offers infinite impedance in dc circuit.

6. *State the factors affecting the capacitance of a capacitor.*
A. Area on plates and distance between the plates. Because $C = \frac{\varepsilon_0 A}{d}$.

7. *What are the applications of R-C time constant?*
A. This is used in high pass and low pass filters in ac circuits.

8. *What is a resistor?*
A. A resistor is an electrical component which is employed to control the current in electrical circuit.

9. *What is a capacitor?*
A. A capacitor is an electrical device or arrangement which comprises of two metallic conductors separated by an insulating medium and carrying equal and opposite charges.

10. *What is the difference between a conductor and a capacitor?*
A. A conductor has some capacitance to store the charge. This quantity of charge is very small. Where as a capacitor has a large capacity to store the charge, because it comprises more than one conductor.

11. *How the reactance of the inductor and capacitor varies with applied frequency.*
A. Inductance allows low frequency and capacitor offer resistance to low frequency.

6. L-C-R Circuit

Aim: To study the series and parallel resonance circuit and to find the resonance frequency, band width and quality factor.

Apparatus: Inductance coil, condenser, a non-inductive resistance box, signal generator, A.C. ammeter, plug key and connecting wires.

Formula: Resonant frequency $(f_r) = \frac{1}{2\pi\sqrt{LC}}$

Band width $= (f_2 - f_1)$

Quality factor $= \frac{f_r}{f_2 - f_1}$

Theory:
Series resonance:

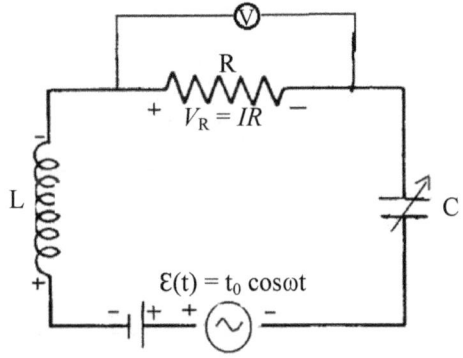

Fig. 7.1 Circuit Diagram

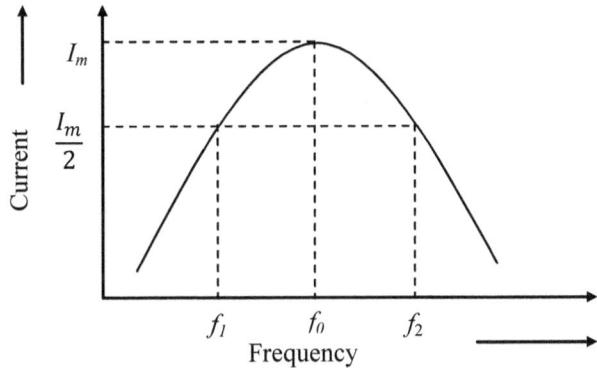

Fig. 7.2 Graph between the frequency and current

The impedance of the series L, C, R circuit is given by
$$z = R + R_0 + j\omega L - \frac{j}{\omega C}$$
Where $\omega = 2\pi f$ and R_0 is resistance of inductance coil.

At certain frequency both reactance becomes equal and this frequency is called resonant frequency and is given by
$$f_r = \frac{1}{2\pi\sqrt{LC}}$$
Here the impedance is minimum and current is maximum

Observation Table:

S. No.	Frequency (KHz)	Current (mA)
1	1.1	
2	1.5	
3	2	
4	2.5	
5	3	
6	3.5	
7	4.5	
8	5	
9	5.5	
10	6	
11	6.5	
12	7	
13	7.5	
14	8	
15	8.5	
16	9	
17	9.5	
18	10	

A graph is drawn between frequency on x-axis and current on y-axis. The frequency at which output current is maximum (I_0) is the resonant frequency (f_r). The frequencies f_1 and f_2 corresponds to $I = 0.707 I_0$.

Then ($f_2 - f_1$) is called bandwidth and $\dfrac{f_r}{f_2 - f_1}$ is called the quality factor (Q) of the circuit.

Parallel resonance:

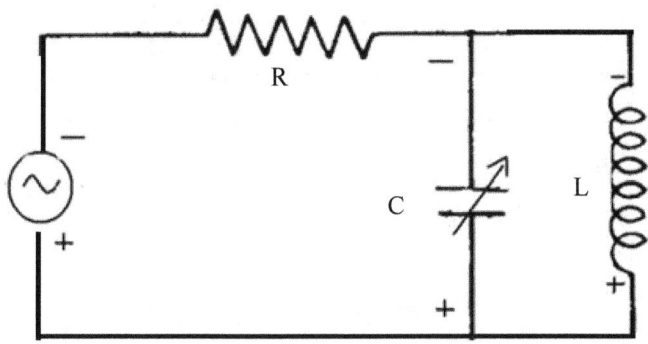

Fig. 7.3 Circuit Daigram

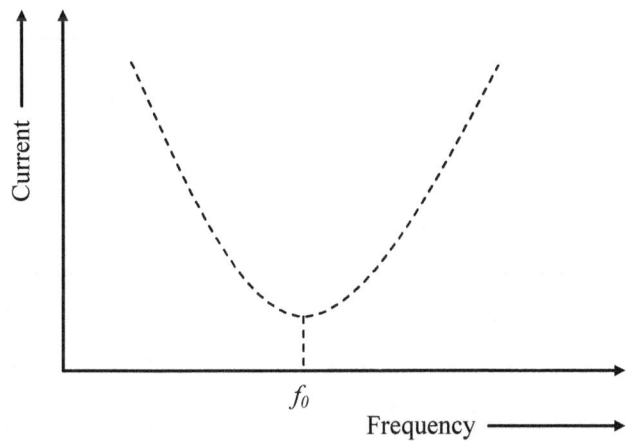

Fig. 7.4 Graph between the frequency and current

The impedance of the parallel L-C-R circuit is given by

$$\frac{1}{Z} = \frac{1}{R_0 + j\omega L} + j\omega C$$

$$Z = \frac{L/C}{R_0 + j\left(\omega L - \frac{1}{\omega C}\right)}$$

Here R_0 is very small in comparison with $j\omega L$, neglect it.

Here $f_r = \frac{1}{2\pi\sqrt{LC}}$

But impedance is maximum and current minimum.

S.No.	Frequency (KHz)	Current (mA)
1	1.1	
2	1.5	
3	2	
4	2.5	
5	3	
6	3.5	
7	4.5	
8	5	
9	5.5	
10	6	
11	6.5	
12	7	
13	7.5	
14	8	
15	8.5	
16	9	
17	9.5	
18	10	

A graph is drawn between frequency on x-axis and current on y-axis. The frequency at which output current is maximum (I_0) is the resonant frequency (f_r). The frequencies f_1 and f_2 correspond to

$I = 0.707\ I_0$. Then $(f_2 - f_1)$ is called bandwidth and $\dfrac{f_r}{f_2 - f_1}$ is called the quality factor (Q) of the circuit.

Results:
Series:
    Resonance of LCR Circuit = _____

    Band width of LCR Circuit = _____

    Quality factor of LCR Circuit = _____

Parallel:
    Resonance of LCR Circuit = _____

    Band width of LCR Circuit = _____

    Quality factor of LCR Circuit = _____

Viva questions

1. What will happen if both capacitor and inductor are connected in a circuit?
A. When a capacitor and an inductor are combined in a single circuit the energy can be traded back and forth between them at any given time. This leads to oscillations in the circuit.
Example: radio receiver

2. What is quality factor?
A. It is the ratio between resonant frequency to the band width.
    i.e., $Q = f_0 / \Delta f$ Where $\Delta f = f_2 - f_1$

3. What is impedance?
A. It is the ratio between maximum voltage values to the current value.
$Z = V_{max} / I$.

4. What is resonant frequency?
A. The maximum / minimum current occurs at particular frequency in the circuit is known as resonant frequency.

5. **What is the relation between impedance and current in this series circuit?**
A. Inversely proportional because in LCR series circuit the impedance is minimum at resonant but we get maximum current at that particular resonant. That means if the current is less the impedance is high and if the current is more the impedance is less.

6. **What is he status of a current in series and parallel connections?**
A. In series the current will be maximum and in parallel the current will be minimum.

Explain the following terms.
i) *Quality factor of a circuit:* The quality factor is a measure of the efficiency of energy stored in an inductor or capacitor when an alternating current is applied. This is defined as 2π times the ratio of the energy stored to the average energy loss per period.

$$Q = 2\pi \; \frac{energy\ stored}{energy\ loss\ per\ period}$$

$$Q = 2\pi f \times \frac{\frac{1}{2}Li_0^2}{\frac{1}{2}i_0^2 R} = \frac{2\pi f L}{R} = \omega \frac{L}{R}$$

Quality factor may also be defined as the ratio of reactance of either inductance or capacitance at the resonant frequency to the circuit. The sharpness of resonance is defined as the rapidity with which the current falls from its value ($^{\epsilon_0}/_R$) with change in applied frequency.

ii) *Band width:* The difference of two half power frequencies that is, $(f_2 - f_1)$ is called as band width of the resonance curve.

$$\text{Band width} = f_2 - f_1 = (f_0/Q) = \frac{f_0 R}{w_0 L} = \frac{R}{2\pi L}$$

Series resonant circuit:
a) Series resonant frequency is given by $f_r = \frac{1}{2\pi\sqrt{LC}}$
b) At resonant frequency, the power factor is unity and impedance is purely resistor $Z_r = R$.
c) At resonance, the current is maximum and the impedance of the circuit is minimum.

d) At the circuit is called as acceptor because it accepts a particular frequency and rejects all others.
e) At resonance the circuit exhibits the voltage magnification and it is equal to quality factor.

Parallel resonant circuit:
a) Parallel resonance frequency is given by $f_r = \frac{1}{2\pi}\sqrt{\frac{1}{LC} - \frac{R^2}{L^2}}$.
b) Power factor is also unity but the impedance is $Z_r = \frac{L}{CR}$.
c) At the resonance the current is minimum and the impedance of the circuit is maximum.
d) The circuit is called as rejecter because it rejects only one frequency.
e) At resonance the circuit exhibits current magnification and it is equal to quality factor.

- The maximum current occurs at a particular frequency called as resonant frequency.
- The peak of the curve depends on the resistance of the circuit, where R is low; the peak is high and vice versa. The peak is known as the sharpness of the resonance. If the high of the peak is more, then sharper is the resonance.
- The series resonant circuit is sometimes called as acceptor circuit, the reason is that impedance of the circuit is minimum at resonance and due to this fact it readily accepts that current out of the many currents whose frequency is equal to its resonant frequency.

7. Magnetic field along the axis of a circular coil

Aim: To study the variation of magnetic field along the axis of a circular coil carrying current.

Apparatus: Stewart & Gees type of tangent galvanometer, a battery, plug key, rheostat, ammeter, commutator and connecting wires.

Formula:

$$B = \frac{\mu_0 n i a^2}{2(x^2+a^2)^{3/2}} \text{ Tesla}$$

Description:

The apparatus consists of a circular frame made up of non magnetic substance. All insulated copper wire is wounded on the frame. The ends of the wire are connected to the terminals, and two tapings from the coil are connected to the other two terminals. By selecting a pair of terminals the number of turns used can be changed. The frame is fixed to a long base B at the middle in a vertical plane along the breadth side. The base has leveling screws. A rectangular non-magnetic metal frame is supported on the uprights. The plane of the frame contains the axis of the coil and this frame passes through the circular coil. A magnetic compass like that one used in deflection magnetometer is supported on a movable platform. This platform can be moved on the frame along the axis of the coil. The compass is so arranged that the center of the magnetic needle always lies on the axis of the coil.

The apparatus is arranged so that the plane of the coil is on the magnetic meridian. The frame with compass is kept at the center of the coil and the base is rotated so that the plane of the coil is parallel to the magnetic needle in the compass. The compass is rotated so that the aluminum pointer reads $0^0 - 0^0$. Now the rectangular frame is along East-West directions.

Theory:

When current flows through the coil the magnetic field produced is in perpendicular direction to the plane of the coil. The magnetic needle in the compass is under the influence of two magnetic fields:

'B' due to the coil carrying current given is given by,

$$B = \frac{\mu_0 n i a^2}{2(x^2+a^2)^{3/2}} \text{ Tesla}$$

where

'n' is number of turns
'a' is the radius of the circular coil
'i' is the current in the coil
'x' is the distance
$\mu_0 = 4\Pi \times 10^{-7}$ Henry/meter

And, the earth's magnetic field 'B_e', which are mutually perpendicular. The needle deflects through an angle θ satisfying the tangent law: $B = B_H tan\theta$

The magnetometer is kept at the center of the coil and rotated so that the aluminum pointer reads $0^0 - 0^0$. Two terminals of the coil having proper number of turns are selected and connected to the two opposite terminals of the commutator. A battery, key, ammeter and a rheostat are connected in series with the other two terminals of the coil. The rheostat is adjusted so that the deflection is about 60^0. The ammeter reading 'I' is noted. The two ends of the aluminum pointer are read (θ_1, θ_2). Then the current through the coils reversed using commutator and the two ends of aluminum pointer are read (θ_3, θ_4). The average deflection 'θ' is calculated. The magnetometer is moved towards east in steps of 2 cm each time and the deflections before and after reversal of current are noted, until the deflection falls to 30^0. The experiment is repeated by shifting the magnetometer towards West from the centre of the coil in steps of 2 cm, each time deflections are noted before and after reversal of current.

Fig. 4.1 Stewart & Gees experiment

S.No	Distance from centre of coil(x) In (meters)	Deflections on East					Deflections on West					Average deflection $(\theta_E + \theta_W)/2$	Tanθ
		θ_1	θ_2	θ_3	θ_4	θ_E	θ_1	θ_2	θ_3	θ_4	θ_W		
1	2×10^{-2}												
2	4×10^{-2}												
3	6×10^{-2}					$B=B_e\tan\theta$ (Experimental)					$B=\dfrac{\mu_0 nia^2}{2(x^2+a^2)^{3/2}}$ (Theoretical)		
4	8×10^{-2}												
5	10×10^{-2}												
S.No	Distance from centre of coil(x)												
1	2×10^{-2}												
2	4×10^{-2}												
3	6×10^{-2}												
4	8×10^{-2}												
5	10×10^{-2}												

The graph drawn between the distance (x) and the magnetic field (or) tan θ gives the variation of the magnetic field.

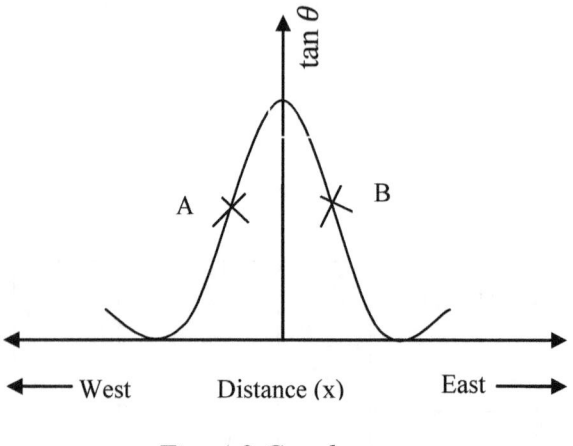

Fig. 4.2 Graph

Result:

Precautions:
1. Galvanometer should not be disturbed while making primary adjustments.
2. Ferromagnetic materials must be kept away.
3. All the measurements must be in SI units.

Viva questions

Explain the following terms.

1. *Deflection magnetometer:* It consists of a compass box and graduated wooden plank. The compass box consists of a short magnetic needle pivoted on frictionless point support as its centre so as to rotate freely in a horizontal plane. The center of the needle coincides with the center of the circular scale graduated in degree and divided into four quadrants. A long aluminium pointer is attached perpendicular to the magnetic needle at its center. The aluminium pointer rotates along with the magnetic needle over the circular scale. A plane mirror is placed below the pointer which helps to note the deflection without parallax error.

The system is enclosed in an evacuated brass case with a glass top. This compass box is placed at the center of a wooden plank. The graduated wooden plank on both sides of the compass box permits the measurements of distances from the centre of the magnetometer box. The two sides of the plank with respect to the compass are known as the arms of the deflection magnetometer. It works on the principle of tangent law.

2. *Tangent law:* Let B and B_H be the two magnetic fields of induction which are perpendicular to each other. When a freely suspended magnetic needle is placed in these fields the needle comes to rest in the resultant direction. If the needle makes an angle 'θ' with the field of magnetic induction B_H, then

$$Tan\ \theta = B/B_H,\ B = B_H\ tan\ \theta$$

This is called tangent law. That is when a magnetic needle is acted upon by two magnetic fields at right angles to each other; the needle will be deflected through an angel 'θ'. Such that the tangent of the angle of deflection is equal to the ratio of the fields.

3. *Fleming's right hand rule:*
Fleming's right-hand rule shows the direction of induced current when a conductor moves in a magnetic field. The right hand is held with the thumb, first finger and second finger mutually perpendicular to each other (at right angles).
- The Thumb represents the direction of Motion of the conductor.
- The First finger represents the direction of the Field. (north to south)
- The Second finger represents the direction of the induced or generated Current (the direction of the induced current will be the direction of conventional current; from positive to negative).

4. *Fleming's left hand rule:*

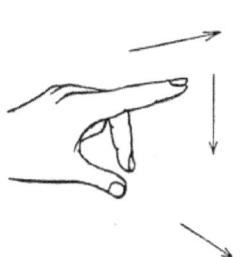

Fleming's left hand rule helps us to predict the movement.
- **First** finger - direction of magnetic **f**ield (N-S)
- Se**C**ond finger - direction of **c**urrent (positive to negative)
- Thu**M**b - **m**ovements of the wire

43

8. Study the characteristics of LED and LASER sources.

Characteristics of LED

Aim: To study the voltage-current characteristics of a given LED.
Apparatus: Regulated power supply (0-30V), voltmeter, ammeter, resistors, LED, connecting wires and bread board.
Theory:

A light emitting diode (LED) is a diode that gives the visible light when forward biased. LEDs are made by elements like gallium, phosphorus and arsenic. By varying these quantities, it is possible to produce light of different wavelengths with colours that include red, yellow and blue.

When LED is forward biased, the electrons from the N-type material cross the P-N junction and recombine with holes in the P-type material. These free electrons are in the conduction band and at higher energy level than the holes in the valence band. When recombination takes place, the recombining electrons release energy in the form of heat and light. The light energy is sufficient to produce quite intense visible light.

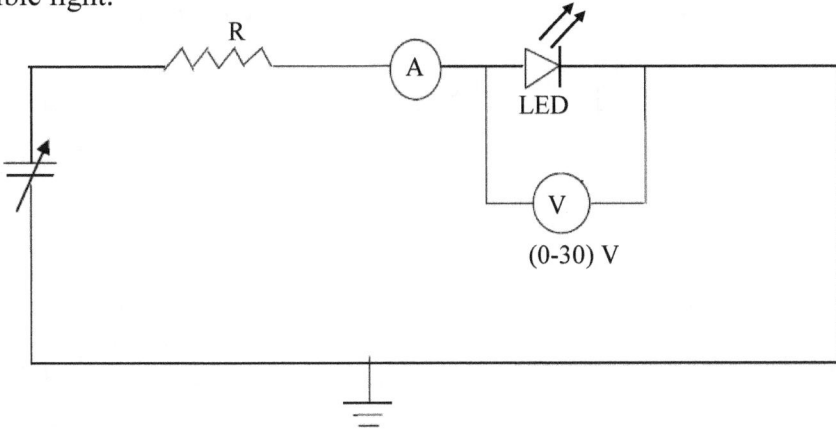

Fig. 8.1 Circuit diagram

Procedure:
1. Connect the circuit as per the circuit diagram as shown in fig. 8.1.
2. Use CQ124 LED and make it forward bias connection.

3. Vary the power supply voltage in such a way that the readings are taken in steps of 0.1 V in the voltmeter till the needle of the power supply shows 20 V.
4. Note the corresponding ammeter readings.
5. Plot the graph between voltage and current.

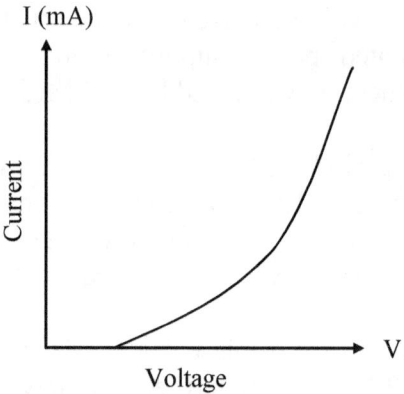

Fig. 8.2 Graph between the voltage and the current

Observations:

SNo.	Voltage (V)	Current (mA)

Precautions:
1. Do not connect the ammeter across the supply (or) to diode.
2. Do not connect the voltmeter in series with the diode.
3. Select the meters of proper range which are somewhat greater than required ratings.

Result:

Characteristics of a LASER diode

Aim: To study the voltage-current characteristics of a given laser diode.
Apparatus: Regulated power supply (0-30V), voltmeter, ammeter, resistors, LASER, connecting wires, bread board.
Theory:

To obtain a laser action in a semiconductor, the medium should be prepared in a form a p-n junction diode with highly degenerate p- type and n-type region, in this way the inverse is produced in the junction region. This can be achieved by forward biasing the junction. When the junction is forward biased with a voltage that is nearly equal to the energy gap voltage, electron and holes are injected across the junction in sufficient number to create a population inversion in a narrow zone called the active region. The amount of population inversion, and hence the gain is determined by the current flowing in the laser diode. At low current values the losses offset lasing action. In this case the radiation exists due to spontaneous emission which increases linearly with the drive current. Beyond a critical value of the current (the threshold value), the lasing commences and the radiative out p_{out} increases rapidly with increasing current.

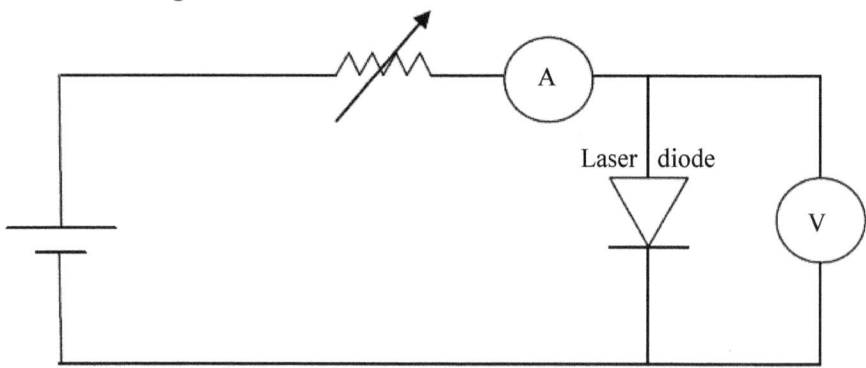

Fig. 8.3 Circuit diagram

Procedure:
1. Connect the circuit as per the circuit diagram as shown in fig. 8.3.
2. Use a laser diode and make it forward bias connection.
3. Vary the power supply voltage in such a way that the readings are taken in steps of 0.1 V in the voltmeter till the needle of the power supply shows 20 V.
4. Note the corresponding ammeter readings.
6. Plot the graph between voltage and current.

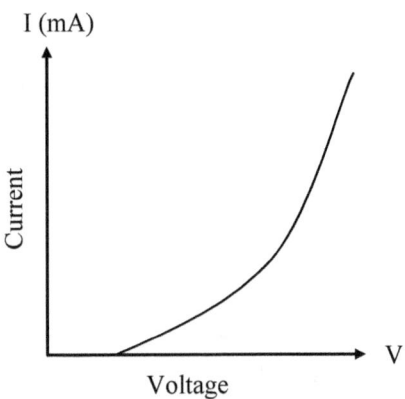

Fig. 8.2 Graph between the voltage and the current

Observations:

SNo.	Voltage (V)	Current (mA)

Precautions:
1. Do not connect the ammeter across the supply (or) to diode.
2. Do not connect the voltmeter in series with the diode.
3. Select the meters of proper range which are somewhat greater than required ratings.

Result:

Viva questions

1. What is a LED?
A. A light-emitting diode (LED) is a semiconductor light source.

2. How a LED is used?
A. LEDs are used as indicator lamps in many devices, and are increasingly used for lighting.

3. What is electroluminescence in a LED?
A. When a light-emitting diode is forward biased (switched on), electrons are able to recombine with electron holes within the device, releasing energy in the form of photons. This effect is called electroluminescence.

4. What are advantages of LED over incandescent light sources?
A. LEDs present many advantages over incandescent light sources including lower energy consumption, longer lifetime, improved robustness, smaller size, faster switching, and greater durability and reliability.

5. What are the practical uses of LED?
A. The first commercial LEDs were commonly used as replacements for incandescent and neon indicator lamps, and in seven-segment displays, first in expensive equipment such as laboratory and electronics test equipment, then later in such appliances as TVs, radios, telephones, calculators, and even watches.

6. What Technology is used in LED?
A. Like a normal diode, the LED consists of a chip of semi conducting material doped with impurities to create a p-n junction. As in other diodes, current flows easily from the p-side, or anode, to the n-side, or cathode, but not in the reverse direction. Charge carriers-electrons and holes-flow into the junction from electrodes with different voltages. When an electron meets a hole, it falls into a lower energy level, and releases energy in the form of a photon.

7. What are the applications of LED?

A. 1. Visual signals where light goes more or less directly from the source to the human eye, to convey a message or meaning.
2. Illumination where light is reflected from objects to give visual response of these objects.
3. Measuring and interacting with processes involving no human vision.
4. Narrow band light sensors where LEDs operate in a reverse-bias mode and respond to incident light, instead of emitting light.

8. What is a photo sensor?
A. Photo sensors or photo detectors are the sensors of light or other electromagnetic energy.

9. What should we do for converting LED to photodiode?
A. LEDs reverse-biased to act as photodiodes.

10. Give examples of photoconductors.
A. Vacuum-tube devices, semiconductor photodiodes, thermocouple semiconductor, photoconductive devices.

11. What is the principle involved in lasing action?
A. Laser is an acronym for Light Amplification by Stimulated Emission of Radiation. The basic principle involved in lasing action is the phenomenon of stimulated emission, which was predicted by Einstein in 1917.

12. What is population inversion?
A. When the no. of atoms are more in higher energy state than the lower energy state, known as population inversion. it is essential for stimulated emission.

13. What is pumping?
A. It is process to achieve population inversion.

14. What is meant by the term coherency?
A. When the light waves are in same phase and with all most same wavelength in light beam known as coherent.

15. What is diffraction?

A. When the light bends or deviates from path due to obstacle known as diffraction.

16. *What is active medium?*
A. The medium in which population inversion is achieved is called active medium.

17. *What is the action of an optical resonator?*
A. It is the combination of two reflecting mirror which is used to increase the intensity of laser light.

18. *What is absemi conductor diode laser?*
A. Semiconductor diode laser is a specially fabricated p-n junction diode. It emits laser light when it is forward biased.

19. *What is LASER?*
A. The term LASER stands for Light Amplification by Stimulated Emission of Radiation. It is a device which produces a powerful, monochromatic collimated beam of light in which the waves are coherent.

20. *What are the characteristic of laser radiation?*
A. Laser radiation has high intensity, high coherence, high monochromacity and high directionality with less divergence.

9. Numerical aperture and bending losses in a given optical fibre

(a) Determination of Numerical Aperture for the given OFC:
Aim: To determine the Numerical aperture and Acceptance angle of the given optical fiber.
Apparatus: Optical fiber cable (1meter), digital multimeter, D.C. power supply, fiber optic trainer module, N.A. jig and N.A. Scale.
Formula: $\quad \text{N.A} = \dfrac{W}{(4L^2+W^2)^{1/2}}$

Description:
The schematic diagram of the fiber optics trainer module is shown in figure

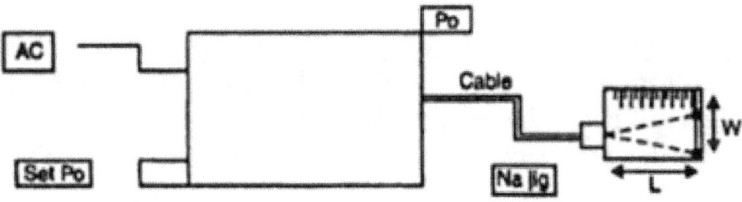

Fig. 9.1 Circuit diagram of OFC

The circuit comprises of three parts:
1. **Electrical to optical (E/O) converter:** It converts an input voltage to an optical output P_0 by driving the fiber optic LED current linearly using a negative feedback operational amplifier circuit. Direct current LED current setting is done by rotating the knob SET/P_0. The optical power is coupled to the optical fiber through the connector. The LED current can be measured by monitoring the voltage with a digital multimeter. A.C. input is given to the V_{in}. $V_r/100$ gives the LED current in milliamperes.

2. **Optical to electrical (O/E) converter:** It accepts the input power (P_{in}) from the optical fiber connected through the connector and provides an output voltage (V_0). Which is directly proportional to P_{in}. For D.C. measurements a multimeter may be used. For A.C. measurements an oscilloscope is required.

3. **Optical power meter:** The optical power meter converts the optical power coupled to it through and SMA terminated optical fiber and facilitates display of power P_0 in dB. The optical power in dB is given by the (multimeter reading/10) in dB referred to $1 mW$.

Theory:
The numerical aperture of an optical system is a measure of the light collected by an optical system. It is the product of the refractive index of the incident medium and the sine of the maximum ray angle.

$$\text{Numerical aperture (N.A)} = n_i \sin \theta_{max}$$

For air $n_i = 1$

$$\therefore N.A. = \sin \theta_{max} \ldots \ldots \ldots \ldots (1)$$

For step-index fiber, the N.A. is given by:

$$N.A = (n_{core}^2 - n_{cladding}^2)^{1/2} \ldots \ldots \ldots \ldots (2)$$

Theory corresponding to experimental arrangement:

Light from the fiber end A falls on the screen BD. Let the diameter of the light falling on the screen = BD = W.

Let the distance between the fiber end and the screen = AO = L.

$$\therefore \text{ from geometry, } AB = \left(L^2 + \frac{W^2}{4}\right)^{1/2} = \frac{(4L^2 + W^2)^{1/2}}{2}$$

$$\therefore N.A = \frac{W}{(4L^2 + W^2)^{1/2}} \ldots \ldots \ldots \ldots (3)$$

Knowing W and L, the N.A can be calculated. Substituting this value of N.A. in equation (1), the acceptance angle θ_A can be calculated.

Procedure:
1. Connect one end of the Fiber optic cable to P_0 and the other end to N.A.jig as shown in fig.
2. The A.C. main is switched ON. Light should appear at the end of the fiber on the N.A. jig to ensure proper coupling is made or not. Turn the SET P_0 knob in clockwise direction to get maximum intensity light through fiber.
3. A screen with concentric circles of known diameter is kept vertically at a distance of 15mm (L) from the fiber end and view Red spot on the screen. Position the screen cum scale to measure the diameter (W) of the spot. Choose the largest diameter.

4. The experiment is repeated for the subsequent diameter of the circles by adjusting the length

S.No.	W (cm)	L (cm)	N.A	$\theta = Sin^{-1}(N.A)$
1	5			
2	10			
3	15			
4	20			
5	25			

Result: Numerical Aperture of given OFC is _____
Acceptance angle of given OFC is _____

(b) Determination of Bending Losses in a given OFC:

Aim: To determine the bending losses in optical fibers
a) Macro bending of the fiber and
b) The connectors between the fibers

Apparatus: One, three & five meters of a step-index optical fiber, digital multimeter, In-line adaptor, , D.C. power supply, fiber optic trainer module

Procedure: Connect one end of the fiber optic cable to P_0 and the other end to the P_{in}.

1. Set the DMM to 2000mV range as the output power is calibrated in terms of millivolts. Connect the two wires of the power output P_0 to the DMM.
2. Plug the A.C. mains. Connect the fiber optic patch cord after relieving all Twists and strains on the fiber. Adjust the SET Po knob a suitable value, say -15 dBm (the DMM will read it as 150mV). Note this as P_{01}.
3. Wind one turn of the fiber on the mandrel in order to remove the twists and strains and note down the reading of the power meter. The difference of these two readings is equal to the loss due to bending.
4. Use the in-line SMA adapter and connect one more cable in series as shown below. Note down the reading P_{02}.
5. Connect one more cable in series to the above two and note down the reading P_{03}.

55

$P_{03} - P_{01}$ = The loss due to the second cable + the loss due to the line adapter.

$P_{03} - P_{02}$ = The loss due to the first cable + the loss due to the line adapter.

Assuming a loss of $1 dB$ in the loss in-line adapter (LILA), we obtain the loss in each cable. The experiment is repeated for next value of P_0.

SNo	$P_{01}(dBm)$	$P_{02}(dBm)$	$P_{03}(dBm)$	Loss in cable 1 $(dB) = P_{03} - P_{02}$	Loss in cable 2 $(dB) = P_{03} - P_{01}$
1					
2					
3					
4					
5					

Result: Bending loss in OFC - 1 is _____
Bending loss in OFC - 2 is _____

Viva questions

1. *What is an optical fibre?*
A. An optical fibre functions as a "light pipe," carrying light generated by lasers and other signal transmission sources to its destination.

2. *What are optical fibre parameters?*
A. Wavelength (λ), core radius (a), index of the core (n_1), index of the cladding (n_2), maximum intensity of light (I_0), numerical aperture (NA), V-number (V), mode type and number of modes.

3. *What is formula for numerical aperture?*
A. Numerical aperture (NA) = $\sqrt{n_1^2 - n_2^2}$

4. *To guide light what should the relationship between n_1 and n_2.*
A. The value of n_1 must be slightly larger than n_2 to guide light.

5. *What are single mode fibres?*
A. Single-mode fibres have a small core size (<10 μm) which permits only one mode or ray of light to be transmitted. Single-mode fibres have low attenuation and zero dispersion at 1310 nm. This fibre is a general-purpose fibre for systems of moderate distance, transmission rates and channel count.

6. *Explain multimode fibre.*
A. Multimode fibres have larger cores that guide many modes or rays simultaneously. When one pulse of a signal is generated into a multimode fibre, the multiple modes enter the fibre core from different angles and each mode propagates at a different speed. This causes pulse broadening (modal dispersion), limiting the speed at which subsequent pulses may be generated without overlapping. Multimode fibres are generally used for short distance applications, such as within buildings.

7. *How fibre-optic transmission works.*
A. The digital bit stream enters the light source. If a one bit is present, the light source pulses light in that time slot, but if there is a zero bit, there is no light pulse (or vice versa, depending on how it is set up). The absence or presence of light therefore represents the discrete ones and zeros. Light energy, like other forms of energy, attenuates as it moves over a distance, so it has to run though amplification or repeating process.

8. *How can we classify transmission medium.*
A. A transmission medium can be classified as (i) linear medium, if different waves at any particular point in the medium can be superposed (ii) bounded medium, if it is finite in extent, otherwise unbounded medium (iii) uniform medium or homogeneous medium, if its physical properties are unchanged at different points (iv) isotropic medium, if its physical properties are the same in different directions.

9. *Define optical medium.*
A. An optical medium is material through which electromagnetic waves propagate. It is a form of transmission medium. The permittivity and permeability of the medium define how electromagnetic waves propagate in it.

10. What are the losses in an optical fibre?
A. Reflection losses, fibre separation, lateral misalignment, angular misalignment, core and cladding diameter mismatch, numerical aperture (NA) mismatch, refractive index profile difference and poor fibre end preparation.

11. Define bending losses.
A. The losses due to the propagation light in an optical fibre (or other waveguide) caused by bending.

12. Explain loss in an optical fibre.
A. Loss is a "relative" power measurement, the difference between the power coupled into a component like a cable or a connector and the power that is transmitted through it. This difference is what we call optical loss and defines the performance of a cable, connector, splice, etc.

13. Explain the scattering in light.
A. The propagation of light through the core of an optical fibre is based on total internal reflection of the light wave. Rough and irregular surfaces, even at the molecular level, can cause light rays to be reflected in random directions. This is called diffuse reflection or scattering, and it is typically characterized by wide variety of reflection angles.

14. Define the numerical aperture.
A. Numerical aperture is defined as the light gathering capability of an optical fibre. It is the sine of the acceptance angle. $NA = \sin \theta_A$

15. What is the principle used in fibre optic communication system?
A. The principle behind the transmission of light wave s in an optical fibre is total internal reflection.

10. Energy gap of a material of P-N junction

Aim: To determine the energy gap of a given semi conductor diode (silicon or germanium).

Apparatus: Semiconductor diode, micro ammeter, voltmeter, thermometer, copper vessel, regulated DC power supply, heater and Bakelite lid.

Formula:
$$E_g = \frac{2 \times 2.303 \times slope \times Boltzmann's\ constant}{1.6 \times 10^{-19}}\ eV$$

Description: The energy gap of a semiconductor is defined as the minimum amount of energy required to an electron to get exited from valence band to conduction band in a given material. The energy gap value is zero for the conductor, it is minimum for the semiconductors and it is maximum in the case of insulator.

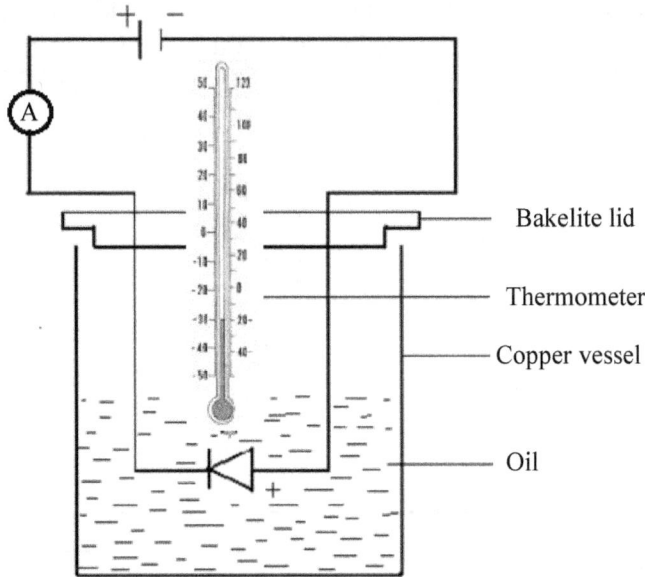

Fig. 10.1 Experimental setup

Procedure: Make the required connections as shown in the circuit. Pour some oil in the copper vessel. Bakelite lid is fixed to the copper vessel a hole is provided on the lid through which thermometer is inserted in to the vessel. Apply the constant voltage to the given semiconductor. When the heater is switched on, measure the current values as the temperature is increasing in steps of 5°C until the temperature is reached to 80°C.

When the heater is switched off, measure the current values as the temperature is decreasing in steps of 5°C until it reaches to the room temperature. This slope is substituted in the given principle to determine the energy gap of a given semiconductor.

Micro board: It consists of Germanium diode (OA 79), thermometer, and copper vessel. Fix the diode to the Bakelite lid such that it is reverse biased. Bakelite lid is fixed to the copper vessel, a hole is provided on the lid through which the thermometer is inserted into the vessel. With the help of heater, heat the copper vessel till temperature reaches up to 80 °C. Note the current reading at 80°C, apply suitable voltage say $1.5V$ (which is kept constant) & note the corresponding current with every 5°C fall of temperature, till the temperature reaches the room temperature.

A graph is plotted between (2.303 $\log_{10} I_s$) and $1/T$ (k) is a straight line. Slop is measured by taking the values of two points where each one of them intersect on the straight line as shown in the figure.

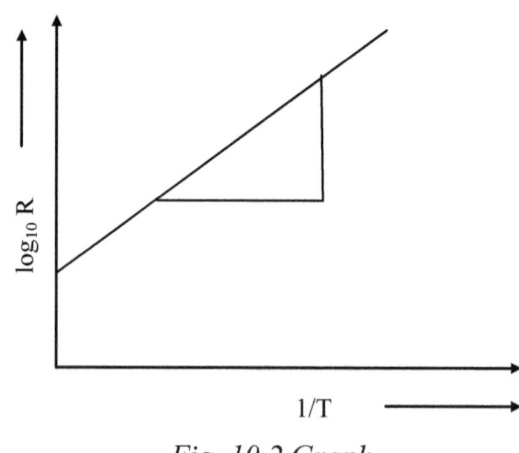

Fig. 10.2 Graph

Applied constant voltage $(V) = 1.5\ eV$.
The Energy gap, E_g = Slope x Boltzmann's constant.
Here, Boltzmann's constant = 1.38×10^{-23} J/K.

$$E_g = \frac{2 \times 2.303 \times slope \times Boltzmann's\ constant}{1.6 \times 10^{-19}}\ eV$$

(1 Joule = 1.6 x 10^{-19} eV)

Depending upon the doping level of the diode the energy gap may vary between 0.5 eV to 0.7 eV.

Note: Do not allow the temperature to rise beyond 100 °C. If you switch off the heater at 80 °C, it will keep on rising for few minutes and may go up to 85^0 to 90^0 before stabilizing/falling.

Observation table:

SNo	Temperature		Current (I)	R=V/I	$Log_{10} R$	1/T
	°C	K				
1	85					
2	80					
3	75					
4	70					
5	65					
6	60					
7	55					
8	50					
9	45					
10	40					
11	35					
12	30					

Result: Energy gap the given semiconductor diode is _____ eV.

Viva questions

Explain the following terms.

1. *Valence band:* The band which consists of valence electrons of an atom or the highest filled band in an electron. The next highest filled band which is partially filled with electrons is known as conduction band.

    The valence band and conduction band are usually separated by certain forbidden energy region. This region is called forbidden energy gap or band.

2. *Insulator:* The substances which do not allow electric current to pass through them. The forbidden energy gap is too large and the concentration of free electros is very small. It requires the energy more than 3ev, only when an electric field of 10^8 volt/meter is applied to an insulator, an electron in valence band can over-come the forbidden gap and can reach conduction band which is practically not possible. Ex: plastic, wood, pvc etc.

3. *Conductor:* Substances that conduct electricity effectively are known as electric conductors. It contains plenty of free electrons. In this the valence band and conduction band overlap each other. The electron can jump to higher energy state with small amount of electric field. The flow of electrons constitutes the electric current through the conductor. Ex: aluminium, copper, silver, iron etc.

4. *Semiconductor:* In an element, if the forbidden energy gap is relatively very small, i.e., about 1ev then that solid is called semiconductor.

    The electrical properties of semiconductor lie between those of insulators and good conductors. The concentration of electrons is approximately 10^{17} electrons per cubic meter. The conductivity of semiconductor is depends on temperature, it increases with increase in temperature because the energy gap of it is decreases with rise in temperature.

    Ex: graphite, pure germanium and silicon.

    Depending upon the relative concentration of charge carriers, semiconductors are again classified as i) Intrinsic and ii) Extrinsic.

5. *Intrinsic semiconductor:* Pure semiconductors are called intrinsic semiconductors. Ex: pure germanium and silicon, the energy gap values of these are $0.785\ eV$ and $1.21 eV$ respectively. It behaves as an insulator at 0 K.

   As the temperature is raised from 0 K to room temperature some of the electrons from valence band acquire thermal energy greater than energy gap value and moves in to the conduction band. As these electrons leaves the vacancies in the valence band then holes will be generated.

   In an intrinsic semiconductor, the number of electrons in the conduction band and the holes in valence band are equal.

6. *Extrinsic semiconductor:* An extrinsic semiconductor is one which contains small amount of impurities doped in to the pure semiconductor.

   Based on the type of impurity, extrinsic semiconductors are again classified as i) p-type and ii) n-type.

7. *P-type semiconductor*: The p-type semiconductor is formed when a small amount of trivalent impurity such as gallium, indium, aluminium, boron etc are added to the pure semiconductor. The conduction of electricity in this semiconductor is mainly due to the flow of holes, hence the majority of charge carriers are holes.

8. *N-type semiconductor*: The n-type semiconductor is formed when a small amount of pentavalent impurities such as arsenic, antimony, phosphorous etc, are added to the pure semiconductor. In this the majority of charge carriers are electrons.

11. Torsional pendulum

Aim: To determine the rigidity modulus of the material of the given wire using a torsional pendulum.

Apparatus: Torsional pendulum, stop watch, meter scale, screw gauge and vernier calipers.

Formula:
Rigidity modulus of the material of the given wire is given by

$$\eta = \frac{4\pi M R^2}{a^4}(l/T^2) \text{ dynes / cm}^2$$

where M = mass of the disk
R = radius of the wire
a = radius of the wire
l = length of the wire
T = time period

Description:
Torsional pendulum consist of a uniform metal disc suspended by a wire whose rigidity modulus is to be determined. The lower end of a wire is gripped in a chuck fixed at the center of the disc and the upper end is gripped in another chuck fixed to a wall bracket.

Theory:
When the disc is turned through a small angle in horizontal plane, it makes oscillations about the axis of the wire. The period of oscillation is given by

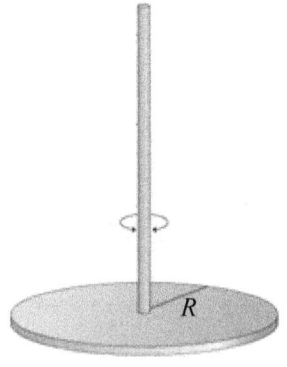

Fig. 2.1 Torsional pendulum

$$T = 2\pi\sqrt{I/C} \quad \ldots\ldots\ldots\ldots\ldots\ldots (1)$$

where I = moment of inertia of the disc about the axis of rotation.
C = couple per unit twist of the wire.

If 'a' is the radius of wire, 'l' is the length of wire between chucks and 'η' is the rigidity modulus of the material of the wire, then the couple 'C' per unit twist of the wire is given by

$$C = \frac{\pi\eta a^4}{2l} \quad \ldots\ldots\ldots\ldots (2)$$

From (1) and (2), we get

$$\eta = \frac{8\pi I}{a^4}(l/T^2)\ldots\ldots\ldots (3)$$

In case of a circular disc whose geometric axis coincides with axis of rotation, the moment of inertia I is given by

$$I = \frac{MR^2}{2}$$

On substituting the value of the value of 'I' in equation (3), we get

$$\eta = \frac{4\pi MR^2}{a^4}(l/T^2) \text{ dyne / cm}^2$$

Procedure:
1. The circular metal disc is suspended as shown in the figure.
2. The length 'l' of the wire between the chucks is adjusted to a convenient value (say 25 cm).
3. The disc is turned in the horizontal plane through a small angle, when it is executing torsional oscillations; time for 10 oscillations is noted twice. Then the time period (T) is calculated.
4. The experiment is repeated for different values of 'l' and results are tabulated in the table.

Observations:

Mass of the disc (M) = _____ gm

Average radius of the disc (R):

Least Count = _____ cm

S. No	MSR	Vernier Coincidence (VC)	Diameter = MSR + (VC x LC)	Radius
1				
2				
3				
		Average radius of the Disc (R)		_____ cm

Average radius of the wire (*a*):
Least count = _____ cm

S.No	PSR	HSR	CHSR	Diameter = PSR + (CHSR * LC)	Radius (a)
1					
2					
3					
				Average	

Readings:

| S. No. | Length of the wire (l) | Time for 20 oscillations (sec) | | | Time period (T) = t/20 | T^2 | l/T^2 |
		Trail I (t_1)	Trail II (t_2)	Mean t= (t_1+t_2)/2			
1	25 cm						
2	30 cm						
3	35 cm						
4	40 cm						
5	45 cm						

Graph:
A graph is drawn taking the value of '*l*' on x-axis and the corresponding values of T^2 on y-axis. It is straight line graph passing through origin.

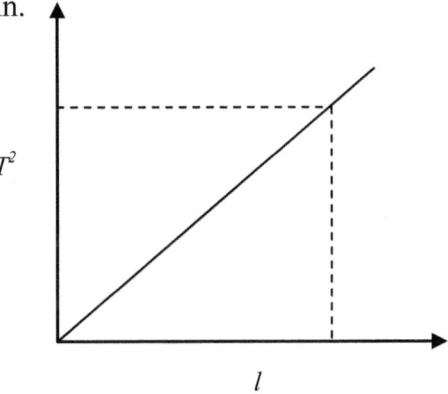

Fig. 2.2 Graph between '*l*' and T^2.

From the graph (l/T^2) is calculated.
Substituting the value of (l/T^2) also, 'η' is calculated.

Precautions:
1. The wire should be free from kings
2. The disc should not wobble.

Result: Rigidity modulus of the material of the given wire (η) is _____ dynes/cm^2

Viva questions

1. *Define the rigidity modulus.*
A. It is the the ratio of tangential stress to the shearing strain. It is denoted by 'η'.

2. *What is the bulk modulus of a material?*
A. The ratio of volume stress to volume strain. It is denoted by 'k'.
   Therefore $k = \frac{F}{A} / \frac{\Delta V}{V}$.

3. *What is the Young's modulus?*
A. The ratio of normal stress to longitudinal strain. It is denoted by 'Y'.
   Y=longitudinal stress/ longitudinal strain, $\frac{F}{A} / \frac{\Delta l}{L}$.

4. *What is the relation between stress and strain?*
A. The stress is directly proportional to strain.
   Stress α strain.
   This is called Hook's law.

5. *What is the elastic limit?*
A. The maximum stress applied on a body up to which it does not get deformed and regains its original position after the removal of applied force.

6. *Explain the stress.*
A. The force applied on a body per unit area. It is again of two types they are,
   1) Normal stress: The stress caused due to the application of force perpendicular to the surface of the body.
   2) Tangential stress: The stress caused in a body due to the application of force in a direction parallel to its surface.

7. *Explain the strain.*
A. The displacement caused in a body due to applied stress. This is again of three types they are,

a) *Longitudinal strain:* The displacement caused in length of a body due to applied force. It is given by $\frac{\Delta l}{L}$.

b) *Volume strain:* The displacement in volume of a body due to applied force is given by $\frac{\Delta V}{V}$.

c) *Shearing strain:* When lower surface of the body is kept fixed and tangential stress is applied on the upper surface then the angle through which it displaces (rotates) is known as shearing strain.

8. *Define moment of inertia.*
A. It is the measure of the inertia of a body in rotatory motion. It depends upon the axis of rotation, mass of the body and also on the distribution of the mass about the axis.

9. *What is the meaning in calling this a pendulum?*
A. The disc is making oscillations around a vertical axis passing through its centre of mass and hence the arrangement is called a torsional pendulum.

10. *What is the difference between simple pendulum and torsional pendulum?*
A. In a simple pendulum the Simple harmonic motion is due to the restoring force which is the component of the weight of the bob.
In a torsional pendulum the Simple harmonic motion is due to the restoring couple arising out of torsion and shearing strain.

11. *What is S.H.M?*
A. A body is said to have a S.H.M, if its acceleration is always directed towards a fixed point on its path and is proportional to its displacement from the fixed point.

12. Wavelength of a laser using grating-diffraction grating

Aim: To determine the wavelength of a given source of laser using a plane transmission grating by normal incident method.

Apparatus: Plane diffraction grating, laser source, scale and grating table.

Formula: $\lambda = 2.54 \sin\theta / nN$

Description: A plane diffraction grating consists of glass plates with equidistant fine parallel lines drawn very closely upon it, by means of a diamond point. The number of lines drawn per inch is written on the diffraction grating by the manufacturers.

The laser consists of a mixture in the ratio of about 10:1, placed inside a long narrow discharge tube. The pressure inside the tube is about 1 *mm of Hg*.

The gas system is enclosed between a pair of plane mirror or pair of concave mirror so that a resonator system is formed. One of the mirrors is of very high reflectivity while the other is partially transparent so that energy may be coupled out of the system. The transition of laser beam corresponding to the red light of wavelength 6328 A° is emitted.

Theory: An arrangement consisting of large number of parallel slits of the same width (e) and separated by equal opaque space (d) is known as the "diffraction grating" ($e+d$) is known as the grating element. A grating is constructed by rubbing equidistant parallel lines 'N' ruled on the grating per inch are written over it.

Hence,

$(e+d) = 1" = 2.54$ *cm*

i.e. the grating element $(e+d) = 2.54$ *cm*

The normal critical incidence the condition for obtaining principle maxima is

$(e+d) \sin\theta = \pm \lambda$

$$\lambda = \frac{2.54 \sin\theta}{nN},$$

where, λ is wavelength of light.

'N' Lines per inch on the plane diffraction grating

'n' is order of diffraction light.

Procedure:. Keep the grating in front of the laser beam such that light is incident normally on it. When light of laser falls on the grating the central maxima along with four other lights are seen on the screen. The light next to central maxima is called the first order maxima and the light next to first order is second order maxima.

Now measure the distance between the grating and the screen and tabulate it as "d_1" and the distance between central maxima to first order and then central maxima and second order is "d_2" and it is also tabulated.

Ray Diagram:

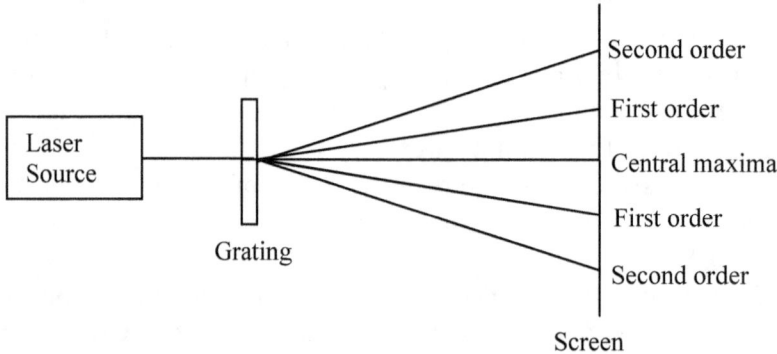

Fig. 11.1 Experimental set up

Observation table:

Number of lines on the grating N = ...
The distance between grating and the screen D = ...

S.No.	Order	LHS (d_1)	RHS (d_2)	Mean (d)	$\sin\theta = d/\sqrt{d^2+D^2}$	$\lambda = \sin\theta / nN$
1						
2						
3						
4						
5						

Precautions:
1. Do not look at the laser beam directly.
2. The prism table should be perpendicular to the laser beam and the grating should be horizontal

Result: The wavelength of a given LASER beam is _____ A^0

Viva questions

i) *What is coherence?*
A. When the electronic transitions taking place in an orderly way and the light waves emitted have a consistent phase relation which doesn't change with time. Because of coherence, tremendous amount of power of the order of 10^{13} watt can be concentrated in a narrow space of linear dimension of $10^{-6}m$.

ii) *Explain the term monochromacity.*
A. The band width of an ordinary laser is of the order of 10 Å and for a high quality laser it is only 10^{-8}Å at 6000Å. This narrow band width of a laser light is called high monochromacity. The spreading of wavelength about the wavelength of maximum intensity is called band width. The phenomena of singe color or single wavelength of laser is known as monochromacity.

iii) *Describe the intensity of light.*
A. Intensity of a wave is the energy per unit time flowing through a unit normal area. The light from an ordinary light source spreads out uniformly in all directions and forms spherical wave fronts around it. If we look at 100 watts lamp filament from a distance of 30 *cm*, the power entering in to the eye is less than 1/1000 of a watt. In case of laser light, energy is emanated in a small region of space and in wave length range and hence is said to be greater intensity.

If we look directly along the beam from a laser, then all the power in the laser would enter the eye. Thus even 1 *watt* of laser would appear many thousand times more intense than 100 *watt* ordinary lamp.

iv) *Define the life time.*
A. The time in which an electron exists in a particular state is known as life time.

13. Characteristics of a solar cell

Aim: To study the characteristics of solar cell.

Apparatus: Two multimeters, variable load, solar cell mounted on wooden base, similar height single directional mercury coated variable intensity source.

Theory: If the depletion of unbiased junction is illuminated, charge separation takes place, resulting in forward bias on the junction. Such device having large area junction very close to the surface is capable of delivering power and is known as solar cell. This cell converts directly solar energy into electricity.

The solar cell radiation is proportional to the delivered power of cell. The efficiency of a cell is expressed in terms of electrical power, output compared with the power in the incident photon flux. The efficiency of solar cell depends on the fraction of light reflected from the surface and the fraction absorbed before reaching the junction. Silicon is widely used for solar cells.

Procedure:
1) Place solar cell directly in front of variable light intensity source and connect output of solar cell to voltmeter 'V' on board.
2) Now gradually increase the intensity of light (bulb) and observe the output of solar cell on the voltmeter 'V'.
3) Then connect the circuit as shown in the circuit diagram [fig 12.1].
4) Vary the intensity, and note voltage and current on 'V' and 'A' meters respectively and as well as connecting load.
5) Plot graph, between voltage and current at different intensities with & without load.

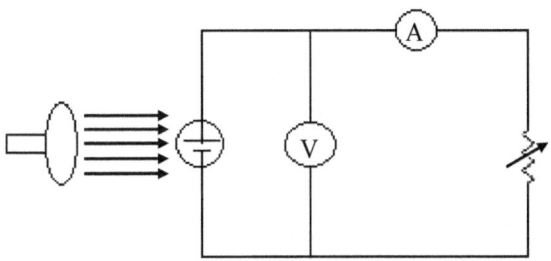

Fig. 12.1 Circuit diagram of solar cell

Observations:

S.No.	Voltage (V)	Current (I)
1		
2		
3		
4		
5		
6		
7		
8		
9		
10		
11		
12		
13		

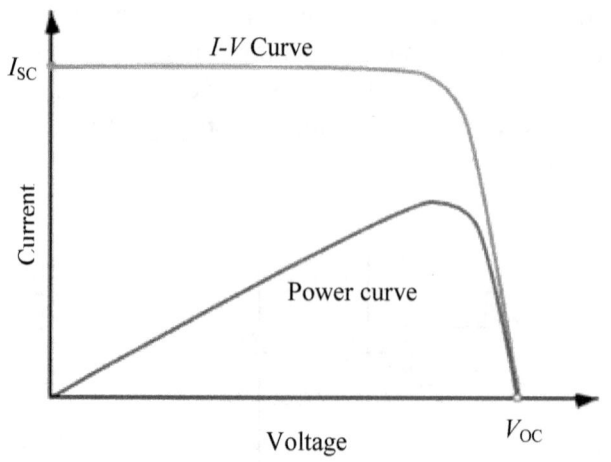

Fig. 12.2 Model graph of I-V curve and power curve

Result: ....

Viva questions

1. *What is a solar cell?*

   A solar cell, or photovoltaic cell, is an electrical device that converts the energy of light directly into electricity by the photovoltaic effect.

2. *How can you find the efficiency of a solar cell.*

   Solar cell efficiency is the ratio of the electrical output of a solar cell to the incident energy in the form of sunlight. The energy conversion efficiency (η) of a solar cell is the percentage of the solar energy to which the cell is exposed that is converted into electrical energy. This is calculated by dividing a cell's power output (in *watt*) at its maximum power point (P_m) by the input light (E, in W/m^2) and the surface area of the solar cell (A_c in m^2).

   $$\eta = \frac{Pm}{E \: x \: Ac}$$

3. *Which types of materials are used in solar cells?*

   Solar cells can be made of only one single layer of light-absorbing material (single-junction) or use multiple physical configurations (multi-junctions) to take advantage of various absorption and charge separation mechanisms. Generally crystalline silicon, cadmium telluride, copper indium gallium selenide etc. are used to prepare the solar cells.

4. *What are the advantages of solar cells?*
   - They have no moving parts and hence require little maintenance and work quite satisfactorily without any focusing device.
   - It does not cause any environmental pollution like the fossil fuels and nuclear power.
   - Solar cells last a longer time and have low running costs.

5. *What is the difference between photodiode and solar cell?*

   A. Functionally both are the same. But there are some differences between them, they are:
   - Size (solar cells are bigger than photodiodes)
   - Power handling capacity (the solar power capacity is more sizeable than photodiodes)

- Application (photodiodes are used as sensors, solar cells are used as transducers which convert light to electricity)

6. *What is short circuit current (I_{SC})?*
A. It is the maximum current from a solar cell and occurs when the voltage across the device is zero.

Additional experiments

14. Hysteresis loop of a ferromagnetic material

Aim: To study the magnetization (*M*) of a ferromagnetic material in the presence of a magnetic field (*B*) and to plot the hysteresis (*M* vs. *B*) curve and to calculate the retentivity and coercivity of the material.

Apparatus: Two solenoid coils *S* and *C*, ferromagnetic specimen rod, reversible key (*R*), ammeter, magnetometer, battery, solenoid, rheostat and transformer for demagnetizing set up.

Theory:

A ferromagnetic rod is magnetized by placing it in the magnetic field of a solenoid. The magnetized rod causes a deflection (θ) in a magnetometer. The deflection (θ) is recorded as the current in the solenoid (*I*) is varied over a range of positive and negative values.

The magnetic field of a solenoid at a point on its axis is

$B = \mu_0 n I$ ------------------------ (1)

where $\mu_0 = 4\pi \times 10^{-7}$ Nm^2/A^2 is the magnetic permeability of vacuum, *n* is the number of turns per unit length in the solenoid and *I* is the current in the solenoid.

The specimen rod is placed along he axis of the solenoid acquires a magnetization *M* along its axis. (Magnetization is defined as the magnetic dipole moment per unit volume).

The magnetic dipole moment '*m*' of the rod is

$m = M(l\alpha)$ ----------------------(2)

where *l* = length of the rod α = cross-sectional area of the rod.

The magnetic field produced by the rod at the position of the magnetometer (*r*) is

$$B_m = \frac{\mu_0}{4\pi} \frac{2mr}{r^2 - \frac{l^2}{4}} \text{------(3)}$$

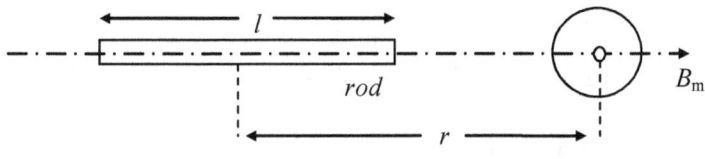

Fig. 8.1

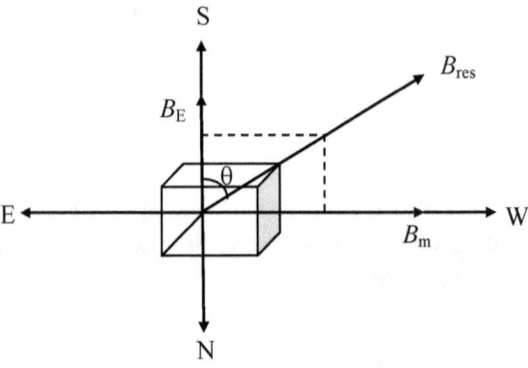

Fig. 8.2

The apparatus is aligned so that the horizontal component of the earth's magnetic field B_E, which is along South- North direction, is perpendicular to the axis of the rod (which is along the East-West direction). The magnetometer needle aligns along the resultant magnetic field making an angle θ with B_E as in fig 8.2.

Clearly

$$Tan\theta = \frac{B_M}{B_E} \Rightarrow B_M = B_E tan\theta \ - - - (4)$$

using equations 2, 3 & 4 we can write

$$M = \frac{4\pi}{\mu_0 a(2l)} \frac{(r^2 - l^2/4)^2}{r} B_E tan\theta \ - - - - (5)$$

Hence $M \propto Tan\ \theta$. Also from eq.1, $B \propto I$. Therefore a plot of $Tan\ \theta$ vs. I reproduces the features of M vs. B curve.

Hysteresis:

A ferromagnetic material whose atoms behave like magnetic dipoles produced by the spins of unpaired electrons. Domains form in the interior of the material with in which the dipoles align in a given direction but the domains themselves randomly oriented. (Fig. 8.3).

In the presence of an external magnetic field the different domain moments tend to align producing a net magnetization in the direction of the magnetic field.

The variation of the magnetization M as the magnetic field B is varied gives rise to a characteristic curve called the hysteresis loop. Figure 8.4 shows a typical curve obtained. (The axes are taken to be $Tan\ \theta$ & I as is to be done in the experiment). As the magnetic field is increased the magnetization of the sample increases as more and more domains align along the direction of the magnetic field. With further increase in B, the magnetization M saturates to a maximum value (point B) If the current I (field B) is decreased the magnetization M decreases.

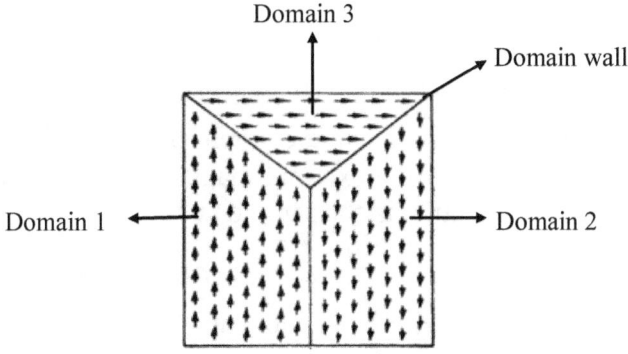

Fig. 8.3 Domains in the ferromagnetic material

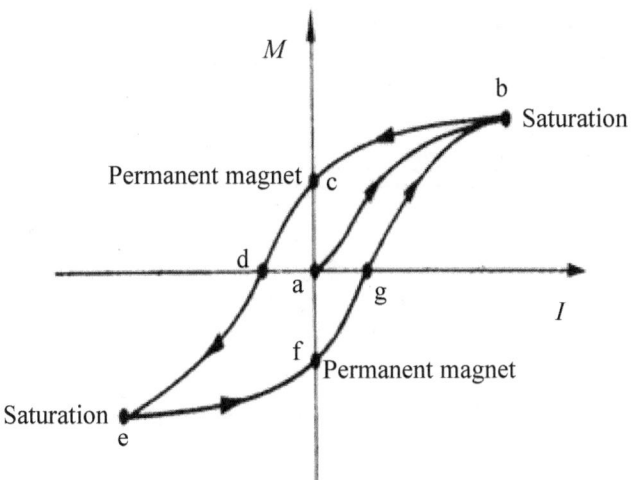

Fig. 8.4. Hysterisys curve

When the current is made zero (point c) the magnetization M however does not fall to zero. At this point the material has a residual magnetization and behaves like a permanent magnet. To make the magnetization zero (point d) requires a non-zero current in the reverse direction. As 'I' is increased in the reverse direction, M saturates to a maximum negative value (point e). Further increase in the current brings the magnetization to zero (point g) and eventually to saturation (point b)

Retentivity and Coercivity:

Retentivity (M_0) is the residual magnetization in the sample when the external magnetic field is zero. This is calculated as

$$M_0 = 4\pi/(\mu_0 a(2l)) \frac{(r^2 - l^2/4)^2}{r} B_E \tan\theta \; ---- (6)$$

where $\tan\theta = \dfrac{cf}{2}$

c and f are the points in the graph, fig 4.-----(7)

Coercivity B_0 is the external magnetic field required to reduce the residual magnetization in the sample to zero.

$$I_0 = \frac{dg}{2}; \; B_0 = \mu_0 n I_0 \; ---- (8)$$

where 'd' and 'g' are the points in the graph.

Setup and Procedure: Complete the wiring of the apparatus according to the circuit diagram shown in the fig.8.5.

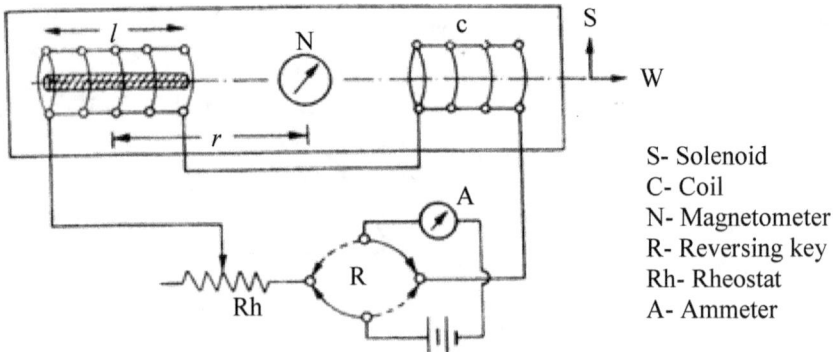

S- Solenoid
C- Coil
N- Magnetometer
R- Reversing key
Rh- Rheostat
A- Ammeter

Fig. 8.5 Experimental setup

Rotate the dial of the magnetometer until 0^0-0^0 position is aligned with the axis of the solenoid. Rotate the wooden arm, containing the solenoid, magnetometer and compensating coil, until the magnetic pointer coincides with the 0^0-0^0 position. In this position the wooden arm is along the E-W position. The horizontal component of earth's magnetic field B_E along S-N direction) is then perpendicular to the wooden arm.

Complete the wiring of the demagnetizing apparatus according to the circuit fig, insert specimen rod in the solenoid and vary the AC current in the solenoid using rheostat. This procedure should take 2-5 minutes.

Pass current (say 1A) through the coils S&C, vary the position of C along the wooden arm until the deflection of the needle is zero. Fig.8.6.

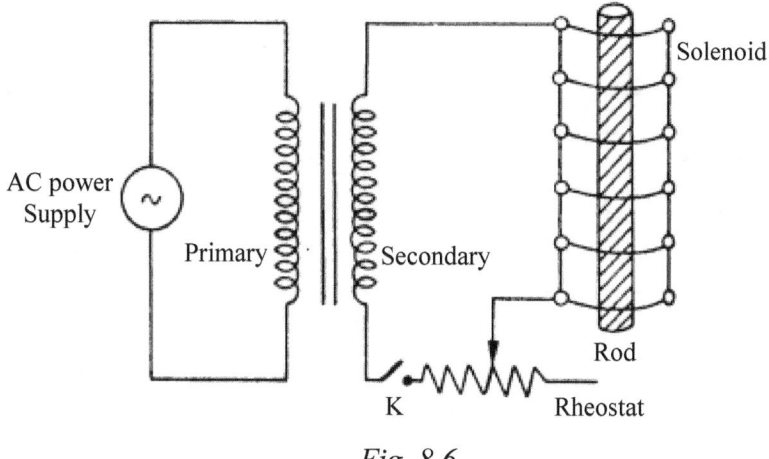

Fig. 8.6

The magnetic field of solenoid S is then nullified (at the position of magnetometer) by the magnetic field of C.

Measurements: to begin with, the current in the solenoid should be switched off. Insert specimen rod so that its leading tip is at the edge of the solenoid. (note: there should be no deflection of the needle at this point. If deflection is observed, repeat for demagnetizing rod).

Keep the reversing key R in a position so that current flows in a given direction. The rheostat position should correspond to maximum resistance.

Switch on the current, vary the current using the rheostat from $0A$-$1.5A$ and back $1.5A$-$0A$ insteps of $0.1A$ and note the deflections $\theta_1 \& \theta_2$ for each setting of current. Note: to get strictly zero current you will have to switch off the battery.

Reverse the position of the reversible key R and vary the current in the reverse direction $0A$-$1.5A$ and back $1.5A$-$0A$. Note the deflections $\theta_1 \& \theta_2$.

Reverse the position of the key R and vary the current from 0-$1.5A$. Again note the deflections $\theta_1 \& \theta_2$.

Observations:

Current I (Amp)	Deflection (forward current) degree				Deflection (reverse current) degree				Deflection (forward current)			
0												
0.1												
0.2												
0.3												
0.4												
0.5												
0.6												
0.7												
0.8												
0.9												
1.0												
1.1												
1.2												
1.3												
1.4												
1.5												
1.4												
1.3												
1.2												
1.1												
1.0												
0.9												
0.8												
0.7												
0.6												
0.5												
0.4												
0.3												
0.2												
0.1												
0												

1) Distance, $r=$ _____ m
2) Length of specimen, $l=$ _____ m.
3) No. of turns per unit length of solenoid, $n=1600$ *turns/m*.
4) Area of cross-section of rod, $S=1.84\times10^{-15} m^2$.
5) Horizontal component of earth's magnetic field, $B_E = 3.53 \times 10^{-5}$ T.

Calculations:
1) Attach graph of $tan\theta$ vs. I_____.
2) Calculation of retentivity:
3) Calculation of coercivity:

Result:
   Retentivity:

   Coercivity:

Viva questions

1. *What is magnetic levitation?*
A. Diamagnetic property of a superconductor ie, rejection of magnetic flux lines is the basis of magnetic levitation. A superconducting material can be suspended in air against the repulsive force from a permanent magnet. This magnetic levitation effect can be used for high speed transportation.

2. *Explain SQUIDS.*
A. SQUIDS (Superconducting Quantum Interference Devices): It is a double junction quantum interferometer. Two Josephson junctions mounted on a superconducting ring forms this interferometer. The SQUIDS are based on the flux quantization in a superconducting ring. Very minute magnetic signals are detected by these SQUID sensors. These are used to study tiny magnetic signals from the brain and heart. SQUID magnetometers are used to detect the paramagnetic response in the liver. This gives the information of iron held in the liver of the body accurately.

3. *What are soft magnetic materials and hard magnetic materials?*
A. *Soft magnetic materials*:

Soft magnetic materials are those for which the hysteresis loops enclose very small area. They are the magnetic materials which cannot be permanently magnetized. In these materials, the domain walls motion occurs easily.

Hard magnetic materials:

Hard magnetic materials are those which are characterized by large hysteresis loop because of which they retain a considerable amount of their magnetic energy after the external magnetic field is switched off.

4. *Define paramagnetic, diamagnetic and ferromagnetic substances.*
A. *Paramagnetic materials*: Paramagnetic materials are those which experience a feeble attractive force when brought near the pole of a magnet. They are attracted towards the stronger parts of magnetic field. Due to the spin and orbital motion of the electrons, the atoms of paramagnetic material posses a net intrinsic permanent moment.

Diamagnetic materials: Diamagnetic materials are those which experience a repelling force when brought near the pole of a strong magnet. In a non uniform magnetic field they are repelled away from stronger parts of the field. In the absence of an external magnetic field, the net magnetic dipole moment over each atom or molecule of a diamagnetic material is zero.

Ferromagnetic materials: These are the ferromagnetic materials in which equal no of opposite spins with different magnitudes such that the orientation of neighbouring spins is in anti-parallel manner are present. Susceptibility positive and large, it is inversely proportional to temperature.

5. *Distinguish soft magnetic material from hard magnetic material in respect of hysteresis losses.*
A. Soft magnetic materials should have low hysteresis losses and low eddy current losses. Easy domain wall motion is the key factor in keeping the hysteresis losses to a minimum. Increasing the electrical resistivity of the magnetic medium reduces eddy current losses. Hard magnetic materials must retain a large part of their magnetization on removal of the applied field. Obstacles to domain wall motion should be provided in permanent magnets so that the energy product high residual induction B_r times larger coercive force H_c is large

15. Volume resonator

Aim: To verify the relation between the volume of air in the resonator and frequency of the note that is in resonance with it.

Apparatus: Aspirator bottle, beaker, measuring jar and forks of different frequencies.

Description: The volume resonator consists of an aspirator bottle with a narrow neck. At its side, there is an opening which is closed by means of a tight fitting rubber stopper, through which passes a glass tube. The glass tube is attached to a rubber tube provided with a pinch clamp. Water can be made to run down through the tube to increase the volume of the enclosed air. The volume of air enclosed can be found by filling the space with water from measuring jar.

Fig. 14.1 Volume resonator

Theory: when a vibrating tuning fork is placed over the neck of the bottle so that the vibrations of the fork are vertical, the air is also set into vibration. When the frequency of the tuning fork is equal to the natural frequency of air inside the resonator, it resonates and a large booming sound is heard. The frequency (n) of note emitted by the resonator is given by

$$n = \frac{\frac{V}{2\pi}\sqrt{A}}{vl}$$

where A = area of neck
l = length of neck

v = velocity of sound in air at room temperature.
V = volume of resonating air
Since A, l and V are constants n^2v = constant.
This is an approximate relation.
The actual relation between n and v is given by
$n^2(v+e)$ = constant, where e = end correction.

Procedure: A tuning fork of known frequency is excited and placed above the neck so that the vibrations are in the vertical plane. The aspirator bottle is filled with water up to the neck. The pinch clamp is opened to run down the water.

The volume of the air enclosed increases and at one particular stage the air enclosed resonates to the frequency of the fork and a large booming sound is heard. The tube is closed and the volume of water collected in a beaker is measured.

Alternatively it is preferred to fill the bottle up to the neck and measure the volume of water required to fill it up to the neck using a measuring jar. This eliminates any error due to spilling of water.

The experiment is repeated using forks of different frequencies. The readings are tabulated as below:

S.No	Frequency of the fork (n)	Resonating volume			$1/n^2$	$n^2(v+e)$
		Trial 1	Trial 2	mean		
1	320					
2	480					
3	512					

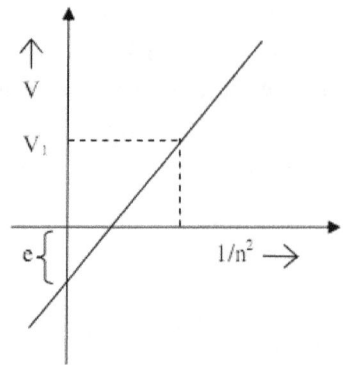

Fig. 14.2. Graph between $1/n^2$ and V

A graph is drawn between $1/n^2$ and v, the volume of air enclosed. It is a straight line. The negative intercept on the y-axis gives the end correction (e). Knowing the value of e and values of $n^2(v+e)$ can be found. They are constants verifying the relation.

Precautions:
1) The excited tuning fork is held above the neck so that the vibrations are in the vertical plane.
2) The aspirator bottle is filled with water up to the neck.

Result:

Viva questions

1. Explain the phenomenon of resonance.
A. Resonance is the tendency of a system to oscillate with greater amplitude at some frequencies than at others. Frequencies at which the response amplitude is a relative maximum are known as the system's resonant frequencies, or resonance frequencies. At these frequencies, even small periodic driving forces can produce large amplitude oscillations, because the system stores vibrational energy.

2. What are the various applications of the phenomenon of resonance?
A.
   i) Acoustic resonances of musical instruments and human vocal cords
   ii) Shattering of a crystal wineglass when exposed to a musical tone of the right pitch (its resonant frequency)
   iii) Electrical resonance of tuned circuits in radios and TVs that allow radio frequencies to be selectively received
   iv) Creation of coherent light by optical resonance in a laser cavity
   v) Orbital resonance as exemplified by some moons of the solar system's gas giants.

3. Explain about the frequency of audible range.
A. Frequency is the number of pressure waves that pass by a reference point per unit time and is measured in Hertz (*Hz*) or cycles per

second. To the human ear, an increase in frequency is perceived as a higher pitched sound, while a decrease in frequency is perceived as a lower pitched sound. Humans generally hear sound waves whose frequencies are between 20 and 20,000 Hz. Below 20 Hz, sounds are referred to as infrasonic, and above 20,000 Hz as ultrasonic.

B. *What is the wavelength of a sound wave?*
A. Wavelength is the distance between two peaks of a sound wave. It is related to frequency because the lower the frequency of the wave, the longer the wavelength.

C. *Describe the amplitude.*
A. Amplitude describes the height of the sound pressure wave or the "loudness" of a sound and is often measured using the decibel (dB) scale. Small variations in amplitude ("short" pressure waves) produce weak or quiet sounds, while large variations ("tall" pressure waves) produce strong or loud sounds.

References:
[1] P. K. Giri, Physics Laboratory Manual for Engineering Undergraduates IIT Guwahati (2005)

[2] B. Srinivasa Rao, V.K. Vamsi Krishna, K.S. Rudramamba Engineering Physics Practicals, Laxmi publications (2013)

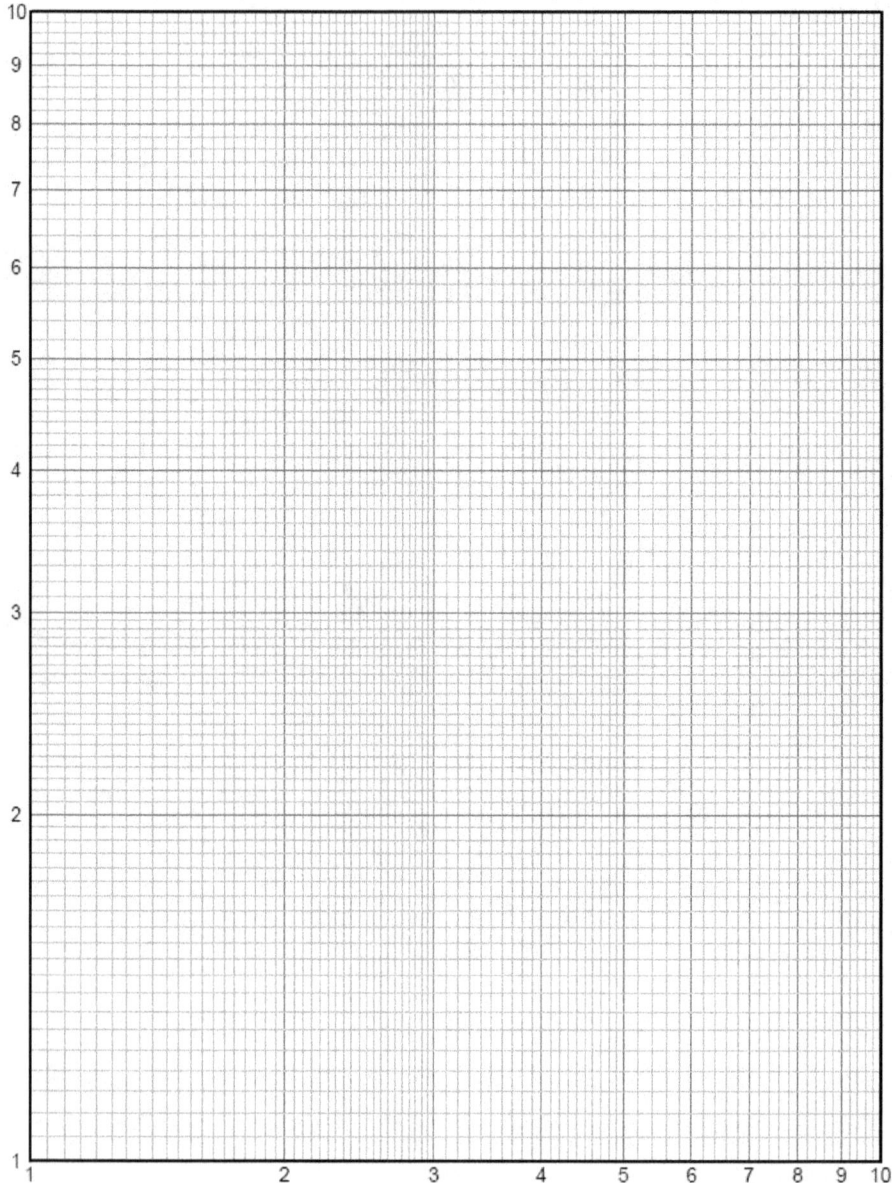

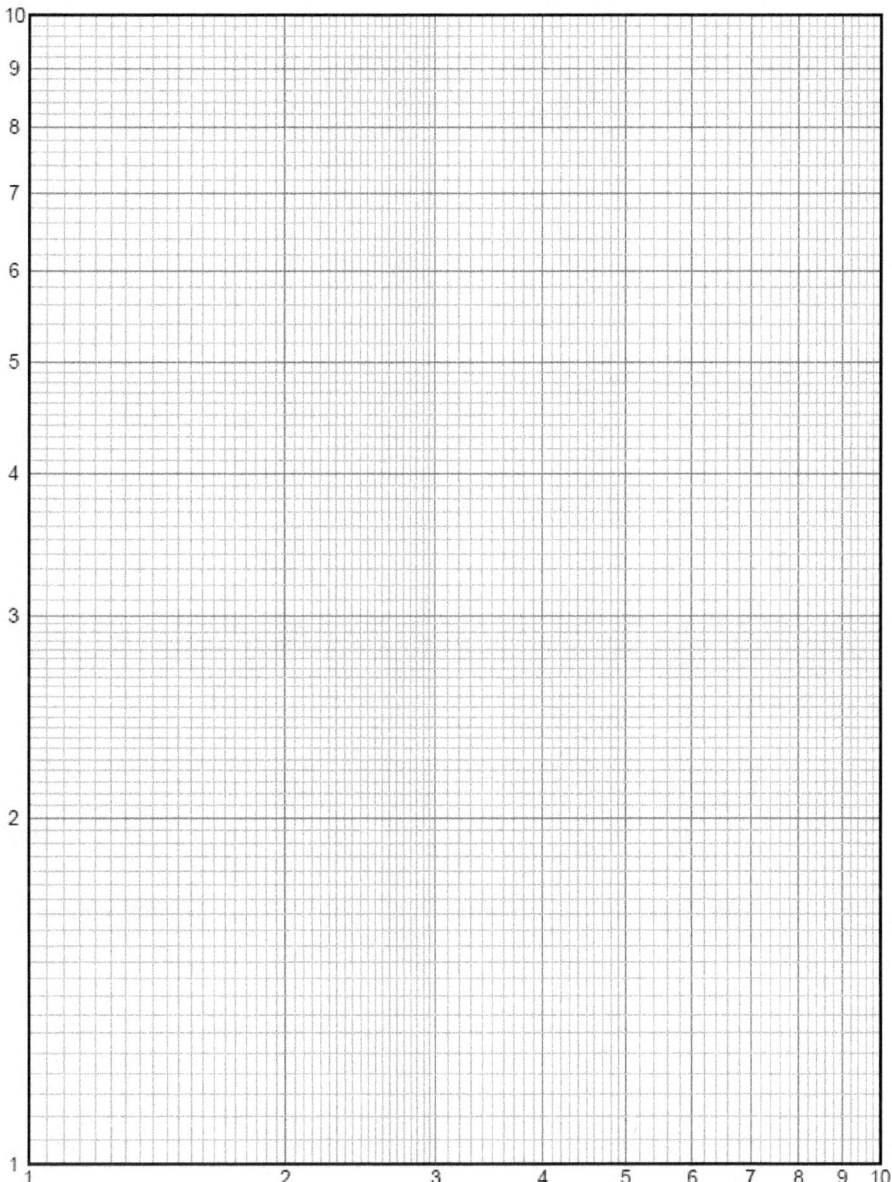

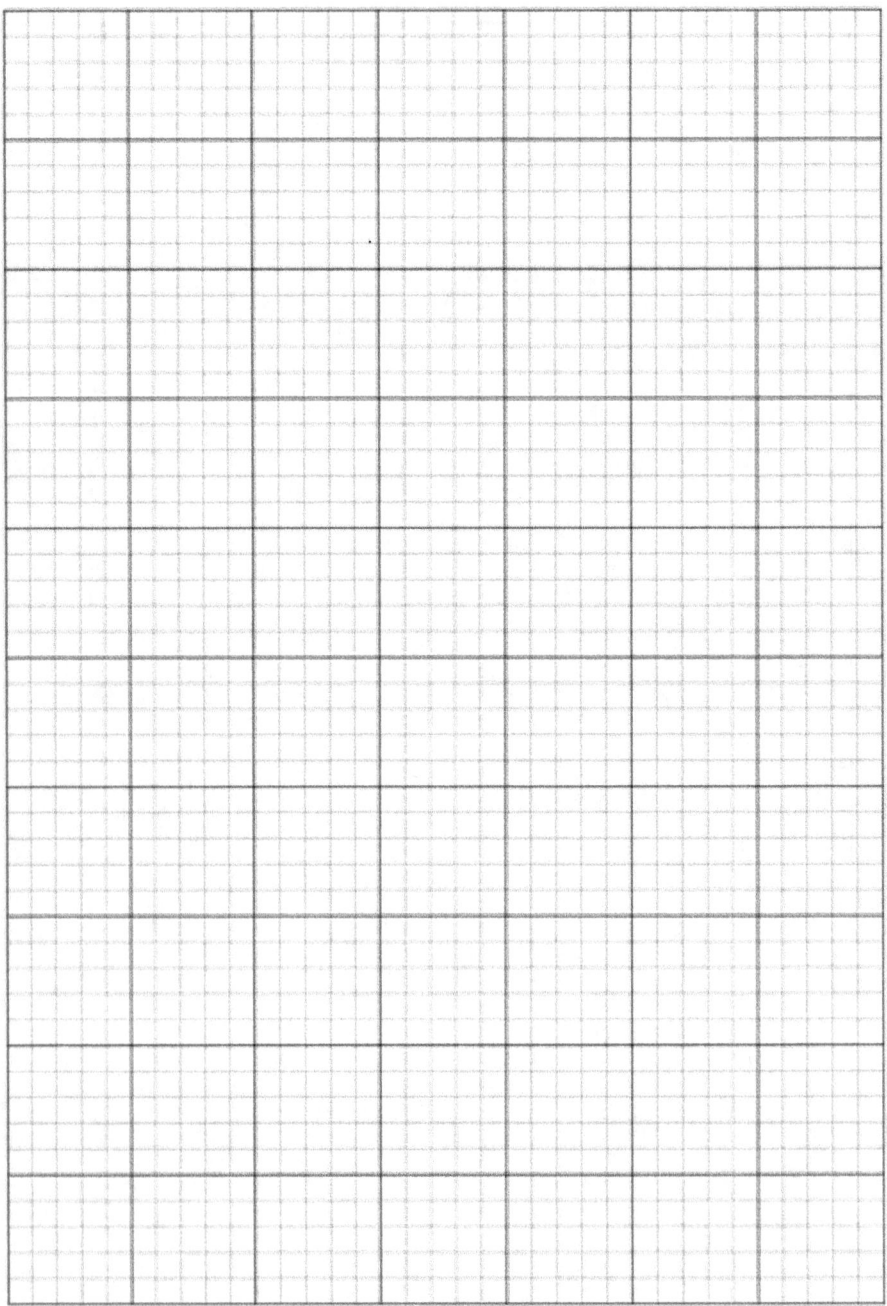

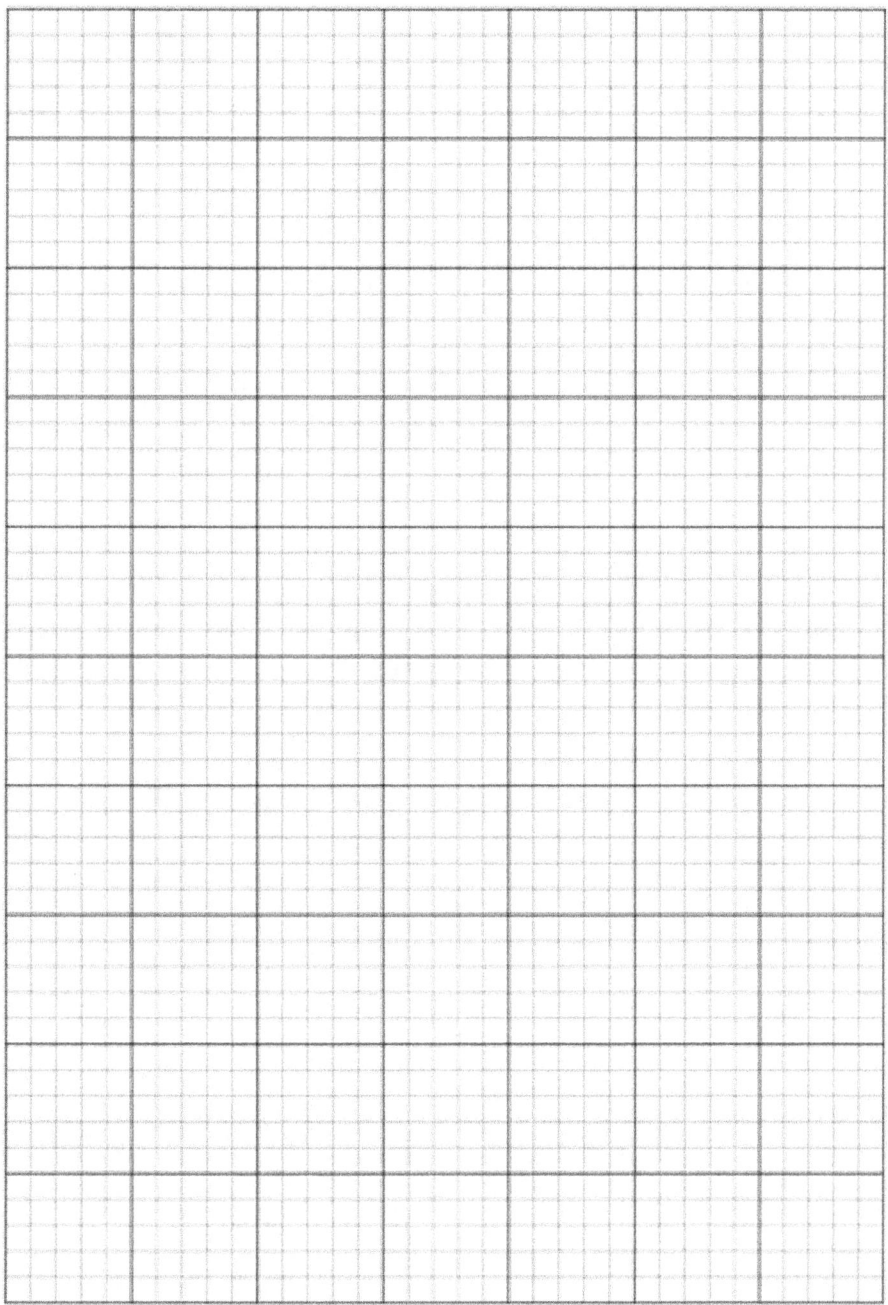

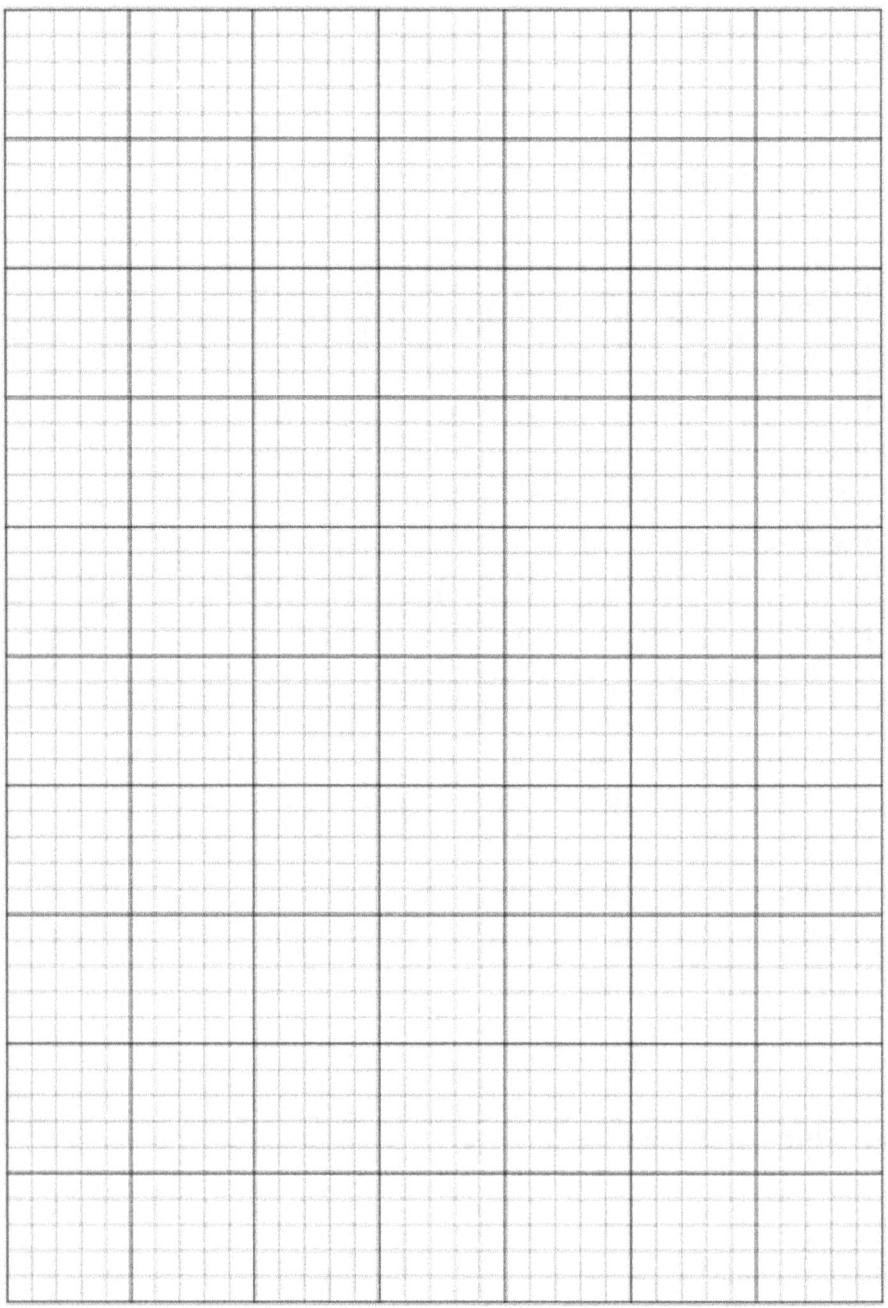

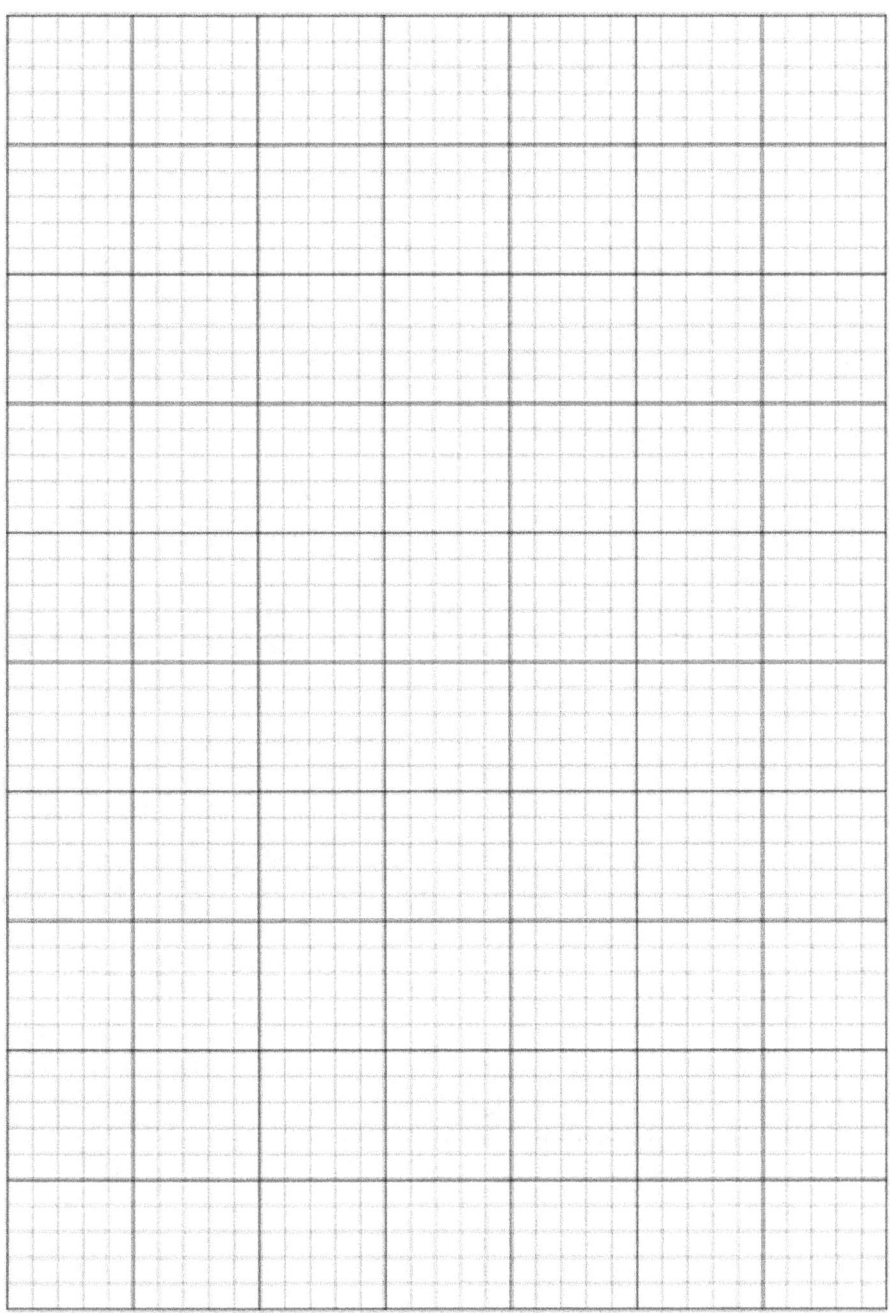

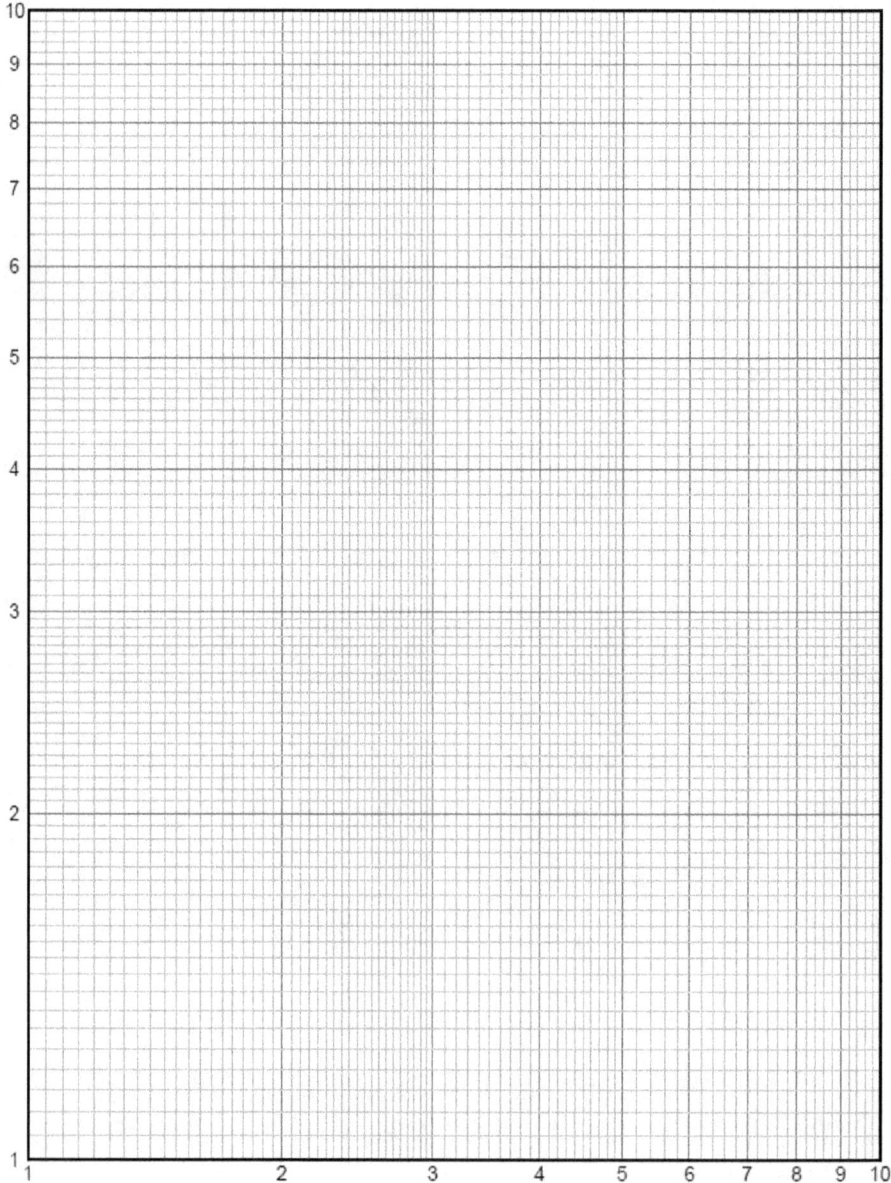

www.ingramcontent.com/pod-product-compliance
Lightning Source LLC
Chambersburg PA
CBHW072222170526
45158CB00002BA/706